Gendered Journeys: Women, Migration and Feminist Psychology

Also by Oliva M. Espín

LATINA REALITIES: Essays on Healing, Migration, and Sexuality

WOMEN CROSSING BOUNDARIES: A Psychology of Immigration and Transformations of Sexuality

LATINA HEALERS: Lives of Power and Tradition

REFUGEE WOMEN AND THEIR MENTAL HEALTH: Shattered Societies, Shattered Lives (*co-author*)

FEMINIST GROWTH AND TRANSFORMATIONS: Asian-American Women in Therapy (*co-author*)

FEMINIST THERAPY WITH LATINA WOMEN: Personal and Social Voices (*co-author*)

SIN OR SALVATION: The Interplay of Sexuality and Spirituality in Psychotherapy (*co-author*)

Gendered Journeys: Women, Migration and Feminist Psychology

Edited by

Oliva M. Espín
San Diego State University, USA

and

Andrea L. Dottolo
Rhode Island College, USA and Women's Studies Research Center, Brandeis University, USA

palgrave
macmillan

First published 2015 by
PALGRAVE MACMILLAN

Palgrave Macmillan in the UK is an imprint of Macmillan Publishers Limited, registered in England, company number 785998, of Houndmills, Basingstoke, Hampshire RG21 6XS.

Palgrave Macmillan in the US is a division of St Martin's Press LLC, 175 Fifth Avenue, New York, NY 10010.

Palgrave Macmillan is the global academic imprint of the above companies and has companies and representatives throughout the world.

Palgrave® and Macmillan® are registered trademarks in the United States, the United Kingdom, Europe and other countries.

ISBN: 978–1–137–52146–0

This book is printed on paper suitable for recycling and made from fully managed and sustained forest sources. Logging, pulping and manufacturing processes are expected to conform to the environmental regulations of the country of origin.

A catalogue record for this book is available from the British Library.

A catalog record for this book is available from the Library of Congress.

To Oliva's grandmothers, whose individual histories of multiple migrations during the nineteenth century defined a particular destiny for her along with her interest in the losses and gains that result from living far from home.

To Andrea's ancestors from Italy, many of whom she never met, and to their descendants, who are with her always, in ways she is still discovering.

To all immigrant women who have struggled and will struggle, thrive and persevere. And to their descendants, who have and will negotiate and integrate their pasts into a livable present and future.

Contents

Acknowledgments

This book is a collective work in more than one way. It would have never come to light without the contribution of all the chapter authors. We start by thanking them for their scholarship and for sharing their stories and experiences in these pages. We thank them for their dedication and their persistence that have made the creation of this collection possible.

Many other people have also contributed to making this book a reality.

Linda Beckman originally encouraged us to write again about immigrant women from feminist psychological perspectives. We thank the anonymous reviewers whose comments strengthened our introduction and proposal and clarified our thinking. Their belief in this work sustained and encouraged us. Susan Cayleff, our colleague and friend, has always been supportive of our academic and personal endeavors. Our past and present students have taught us much. Their questions and challenges have been an inspiration throughout our teaching careers.

I (Oliva) want to thank Andrea for the many long and helpful conversations and her unwavering friendship. Her collaboration and energy through the life of this undertaking have been invaluable. Her faith in this project strengthened my own.

The Women's Studies Department at San Diego State University has been my academic home for 25 years. My colleagues in the department, past and present, as well as colleagues from the California School of Professional Psychology of Alliant International University in San Diego and San Francisco, respectively, have provided stimulating conversations and challenging opportunities.

Members of my writers' group – Sheryl Tempchin, Lillian Faderman, Janice Steinberg, Anne Marie Welsh, Carolyn Marsden and Abigail Padgett – have encouraged me over several years to write about memory and language.

Nery, my sister, has shared many life experiences with me and knows my history better than anyone else. I have learned from the migration experiences and struggles of many women, such as Virginia Aponte, Esther Rothblum, Lourdes Rodríguez-Nogués, Raquel Matas and Krizia Puig. They have been by my side and cheered me on at different points in my life. Many of my therapy clients and other women immigrants I have met through the years have also increased my understanding of what migration entails for women. My stalwart friends Natalie Porter and Ellyn Kaschak have encouraged my work through several decades. Students in my courses on women's experience of migration and LGBT issues in migration have helped me refine my thinking on these topics. Finally, I want to thank Jane Via and Nancy Corran, the women priests in my church, Mary Magdalene Apostle Catholic Community, for their strength and courage, their clarity of thought and their support and nourishment of my spiritual and personal journey.

I (Andrea) first and foremost thank Oliva Espín, without whom this book would not be possible. The hours of conversations of collaboration, commiseration and creativity between us on this project continually remind me of how much of my thinking is a product of Oliva's dedication, insight and generosity. There are theories and terms that will forever live in my head with a Cuban accent. She has always insisted that I can and should do this work, even when so many others said I could not. There are no adequate ways to describe what her friendship and mentorship have taught me about what could be imagined for my possible selves.

This project has also traveled with me among different institutional homes where several individuals and groups provided encouraging support. I would like to thank the Social Issues and Narrative Methodologies Research Study Groups, especially Rahel Wasserfall and Ellen Rovner – my fellow "foodies" – at the Women's Studies Research Center at Brandeis University for their incisive comments, enthusiasm and academic support. And to our fearless leader, Shula Reinharz, whose guidance and direction have served as a model for me and so many others. Rhoda Unger has been a kind and thoughtful mentor, whose stories and actions of bold and keen feminist courage instruct and inspire me.

Meg Bond, Julie Nash and Mignon Duffy were compassionate beacons of light and hope, and they had an unexplained faith in me. I want to thank the Psychology Department and the Qualitative Research Group at Rhode Island College for providing me with a new, welcoming and supportive academic home. I will always be grateful to the Women's Studies Department at San Diego State University for the best feminist training I could imagine.

I am also immeasurably grateful to all the participants in my research and their immigrant ancestors. The words of so many Italian American women are threaded through my consciousness and conceptions in my work. The bookshelf next to my office desk contains the writings of many of them that I covet, pet and tend to and that are strategically located in my line of vision for their vision. These include Jennifer Guglielmo, Louise DeSalvo, Kym Ragusa, Joanna Clapps Herman, Annie Lanzillotto, Edvige Giunta and Helen Barolini. The Italian American Psychological Society of the American Psychological Association has also served me food, conviviality and community at their lively preconvention dinners.

Several friends and colleagues continue to push, cheer and teach me, in our conversations and in their example. Erika Gasser feeds my fetish for history, food and raucous laughter. Champika K. Soysa endures unending hours of discussions, generously giving her analysis, patience and friendship. Sarah Tillery is my brave, brilliant, heart friend who has been a critical witness of my scholarship and life for so many years.

My partner, Sandy McEvoy, has traveled with me as companion and support, and has been an integral part in the development of this continuing work from its inception. As I struggle to write these acknowledgments, and tell her so, she grounds me again, saying, "Remember, it's just a draft. Spit it out, and we'll look at it together." She continues to teach me as we learn to apply this lesson in so many contexts – with and alongside each other.

My family, of course, is the center of this work, the reason why I was ever interested in our collective stories of migration and identity in the first place. I am grateful for Emileen Butler, at my mother's kitchen table, cooking up ideas with coffee, love, laughter, lots of yelling and something sweet. To Lucy and John Berardi, who never would have imagined that their names would appear in print, in a book, and now they are. To my sister, Danielle Dottolo, for her humor, strategic gift giving, and our shared history. To my father, Tony Dottolo, for his kindness, strength, and for always telling me I could be anything I wanted. To my mother, Carol Dottolo, my co-author, for her trust in doing this work, her interviews and teaching me to read at a very early age – also symbolic for the ways in which she teaches me to "read," interpret and understand our recipes, stories and identities.

Nicola Jones and Elizabeth Forrest at Palgrave Macmillan were enthusiastic about this book and have guided us to put the final touches to this work.

To all of them go our heartfelt thanks.

Notes on Contributors

Huma Ahmed-Ghosh is Chair and Professor in the Department of Women's Studies at San Diego State University. She is also on the Advisory Board of the Center for Islamic and Arabic Studies and the Center for Asia-Pacific Studies. Her research focuses on women in Afghanistan, Muslim immigrant women to the USA, and Islam and feminism. Ahmed-Ghosh has traveled frequently to Afghanistan, where she was looking at how Afghan women-run non-government organizations strategize women's rights within an Islamic state. She has published extensively on her research on Muslim women's issues in India and in the USA and on women in Afghanistan. Ahmed-Ghosh has also published articles on gendered aging in India, widowhood, domestic violence and women's representations in beauty pageants in India. Currently, she has two books on gender and Islam in Asia under review with SUNY Publishers and is co-authoring a book on how women peacemakers around the world negotiate gendered violence in conflict at the grassroots level.

Sundari Balan is a post-doctoral research scholar at the School of Medicine of Washington University in St. Louis. She studies how psychological factors and socio-cultural context intersect in relation to well-being, mental health and addictive behaviors. Her own experiences as an Indian mother of two children inspired her dissertation and the chapter presented in this book.

Peggy Brady-Amoon is a faculty member in the Department of Professional Psychology and Family Therapy, Seton Hall University in New Jersey, after spending nearly 30 years as a counselor and counselor supervisor, primarily with college opportunity programs that admit, retain and graduate at-risk students. She has a doctorate in Counseling Psychology. Her research is broadly focused on educational and career development, with a particular emphasis on access and opportunity for under-represented groups, including immigrants and the children of immigrants.

Alison Cerezo is an assistant professor in the Department of Counseling at San Francisco State University. Her scholarly and community initiatives center on higher educational access and persistence for underrepresented students as well as LGBT immigration.

Karen Kisiel Dion is interested in the social psychology of immigration. This includes the distinctive contribution of gender to the experience of immigration. She has done interdisciplinary research collaboration resulting in a co-authored book on multiculturalism and other publications.

Roxanne A. Donovan is Associate Professor of Psychology and African and African diaspora studies at Kennesaw State University in Georgia. Her scholarship examines how the intersections of socio-demographic variables such as gender, racialization and immigration impact the mental health and well-being of people of color, striving to make visible the influence of access and power on those whose voices are typically heard, valued and normalized and whose are typically silenced, ignored and pathologized. She has published on the contemporary stereotypical images of black women and their influence on black women's self-perception and behavior; the impact of gendered racism and racial macro- and micro-aggressions on depressive and anxious symptoms in black women; and how identity, ethnicity and generational status influence depressive symptoms in college students. Donovan is licensed as a psychologist in Georgia.

Andrea L. Dottolo is Assistant Professor of Psychology at Rhode Island College and Resident Scholar at the Women's Studies Research Center at Brandeis University. Her scholarship explores how social identities are constructed and maintained and the ways in which they are shaped by social and political structures. She is interested in the relationships between histories, cultures, social movements, immigration and social identities. Dottolo has a joint PhD in psychology and women's studies from the University of Michigan.

Carol Dottolo has an MA in literacy and taught primary school in the Liverpool, NY school district, for 25 years. She is a reading specialist, and she has extensive experiences in teaching language and comprehension and in developing innovative pedagogical strategies in reading and writing. As the daughter of an immigrant father, she was born and raised in Syracuse, NY, in an Italian American immigrant culture. She is interested in food, recipes, tradition and the intergenerational transmission of culture.

Oliva M. Espín is Professor Emerita in the Department of Women's Studies at San Diego State University and the California School of Professional Psychology of Alliant International University. A native of Cuba, she received her BA in Psychology from the University of Costa

Rica and her PhD from the University of Florida, specializing in counseling and therapy with women from different cultures and in Latin American studies. She did post-doctoral work at Harvard University with a fellowship from the National Institute of Mental Health. Among many professional awards, she has received the American Psychological Association Award for Distinguished Professional Contribution to Public Service and the Distinguished Career Award from the Association for Women in Psychology. The APA Division of International Psychology honored her in for her outstanding international contributions to the psychology of women and gender and named her outstanding international psychologist. Throughout her career she has done research, teaching and consultation and has published many articles and books on immigrant women, including *Latina Realities: Essays on Healing, Migration, and Sexuality* and *Women Crossing Boundaries: A Psychology of Immigration and Transformations of Sexuality*. Espín held the Fulbright Distinguished Chair in gender studies at the University of Klagenfurt, Austria in 2010.

Diya Kallivayalil is staff psychologist at the Cambridge Health Alliance and an instructor in psychology in the Department of Psychiatry at Harvard Medical School. She received her PhD at the University of Illinois at Urbana-Champaign in the Clinical/Community Psychology Division and completed a two-year fellowship at the Victims of Violence Program, Cambridge Health Alliance/Harvard Medical School, specializing in the assessment and treatment of psychological trauma. Her research is broadly concerned with psychological trauma. She has published in the areas of domestic violence, feminist therapy, the impact of trauma and migration on mental health in minority and immigrant communities, narrative methods, political trauma and homicide bereavement. She is the co-author of the *Trauma Recovery Group* (2011) published by Guilford Press.

Pei-Wen Winnie Ma is an associate professor in the Department of Psychology at William Paterson University in Wayne, NJ. She is also a licensed psychologist working part-time at Hamilton-Madison House serving the Asian immigrant population in New York City. Her research investigates career development, family relationships and mental health issues affecting Asian immigrant children, youth and their families. She received her PhD in Counseling Psychology from Teachers College, Columbia University.

R. Lillianne Macias is a graduate research associate for the National Latin@ Research Center on Family and Social Change, a project of Georgia

State University and Casa de Esperanza's National Latin@ Network for Healthy Families and Communities. Macias is a doctoral candidate completing her training in clinical and community psychology under the direction of Julia Perilla.

Ramaswami Mahalingam is Associate professor in the Department of Women's Studies and the Department of Psychology at the University of Michigan. His research focuses on immigration, gender and mindfulness using an intersectionality framework. His current research examines the relationship between intersectional awareness, mindfulness and privilege awareness.

Sandra Mattar is Associate Dean of the Kalmanovitz School of Education, and an associate professor in the Graduate Counseling Program at Saint Mary's College of California. Mattar's research interests are mainly on culturally sensitive education and training, multicultural awareness, the intersection of psychological trauma and culture, immigration, mental health disparities and cultural considerations in disaster mental health. She has presented her work at national and international conferences and has served as an editor on international journals and book collaborations.

Alejandro Morales is an assistant professor in the joint Department of Psychology and Sociology at the California State Polytechnic University, Pomona. His research is in the areas of language brokering in immigrant families, psychological adjustment of immigrant sexual minorities and research methods. He teaches courses on basic counseling skills, intimate relationships and couple's therapy. He received his BA from the California State University, Dominguez Hills, and his MA and PhD from the University of Nebraska–Lincoln.

Neha Patel is a clinical psychologist in independent practice at Chicago, Illinois. She is also a candidate at the Chicago Psychoanalytic Institute.

Julia L. Perilla serves as the director of the National Latin@ Research Center on Family and Social Change, a project of Georgia State University and Casa de Esperanza's National Latin@ Network for Healthy Families and Communities. Perilla is a clinical-community psychologist and faculty member in the Psychology Department at Georgia State University.

María Teresa Pestana graduated with a major in sociology and a minor in political science. Her interests include civil rights, social justice, public interest law, immigration and international relations.

Danielle Quintero is a doctoral student in counseling psychology at the University of Missouri–Columbia. Her research interests center on sexual and domestic violence prevention and on intervention methods within ethnic minority populations. Her current work also integrates social justice and advocacy while examining resiliency, coping and cultural factors for survivors.

Rebecca Rodriguez is a doctoral student in community psychology at Georgia State University, and she conducts research for the National Latin@ Research Center on Family and Social Change under the direction of Julia Perilla.

Alvina Rosales is in the process of completing her PhD in Clinical Psychology at Georgia State University. Under the mentorship of Julia Perilla, Alvina approaches research from empowerment, ecological systems and social justice theoretical perspectives.

Stephanie Rothman is a clinical psychology doctoral student at California School of Professional Psychology in San Francisco. Her research interests involve biopsychosocial and phenomenological-existential analyses with intercultural aging and LGBT populations. She trains as a clinician with geriatric populations and serves as a group facilitator for LGBT elders.

Anmol Satiani is a counseling psychologist in independent practice in Chicago, Illinois. She is also an adjunct instructor at DePaul and Roosevelt Universities and a volunteer therapist at Heartland Alliance Marjorie Kovler Center, a program serving torture survivors in the Chicago area.

Josephine V. Serrata serves as the research manager for the National Latin@ Network for Healthy Families & Communities. Serrata is a clinical-community psychologist who received her doctorate from Georgia State University. She completed her graduate training at Georgia State University.

Munyi Shea is Associate Professor at the Department of Psychology, California State University–Los Angeles. Her research focuses on school experience and psychological adjustment of Asian and Latino immigrant students and culturally responsive prevention and intervention programs. Her research also explores issues related to femininity and women's psychological health outcomes. She received her PhD in Counseling Psychology from Teachers College, Columbia University, and she completed a clinical psychology internship at the Massachusetts Mental Health Center/Harvard Medical School.

Gabrielle Stutman has been in private practice as a licensed psychologist since 1982, with offices in New York City and Ardsley, Westchester. She has advanced training in psychoanalytic psychotherapy, clinical neuropsychology and forensic psychology. While teaching developmental psychology at CUNY (1990–2010) to a largely Latina student population, she became interested in these women's experience of immigration and adaptation to the US. Her training in forensic psychology led her to evaluating families of undocumented immigrants as they attempted to legalize their residential status in this country. Her practice currently includes these forensic evaluations as well as psychotherapy, cognitive remediation and neuropsychological forensic evaluations. An active member of her professional community, she has assumed leadership positions in the New York State Psychological Association since 1997, has four times been president of the Independent Practice Division, is founder of the Forensic and Neuropsychological Divisions and has received several awards for distinguished service. Her publications include articles in the areas of psychoanalysis, neuropsychology and forensic immigration evaluations.

Karen L. Suyemoto is Associate Professor of Psychology and Asian American studies at the University of Massachusetts, Boston. She has conducted research and taught undergraduate and graduate classes focused on racial justice for almost 20 years. In her teaching, scholarship and organizational leadership, she seeks to empower those who have been oppressed or silenced while motivating those with dominant statuses to understand the influence of systemic privilege and work as allies to address injustice. Her scholarship focuses particularly on how racial and ethnic identities and experiences of discrimination are associated with development and mental health for Asian Americans, how individuals and communities create and maintain meanings of race and ethnicity that contribute to or resist oppressive systems and the processes and effects of antiracist and social justice education and community interventions. She has authored over 30 journal articles and book chapters, and she has recently co-authored a book for training therapists: *Conceptualization and Treatment Planning for Effective Helping*. Suyemoto provides consultation and training on antiracist therapy and education both locally and nationally and is the past president of the Asian American Psychological Association.

Pratyusha Tummala-Narra is clinical psychologist and Assistant Professor in Counseling, Developmental and Educational Psychology at Boston College. She is also in independent practice in Cambridge, Massachusetts.

Introduction

Oliva M. Espín and Andrea L. Dottolo

"Migration is a highly complex phenomenon, with significant economic, socio-political, cultural, and religious repercussions for the migrants, their native countries, and the host societies" (Phan, 2003, p. 3). The present book is an effort at understanding this phenomenon of migration from the perspective of the psychology of women and girls.

> The "age of mobility" is here to stay. Never before in human history have so many people been on the move. Today around 214 million people, or approximately 3 percent of the world's population, live outside their country of birth. Women and girls account for half of international migrants, and 16 percent are under the age of 20. South-south migration is now as frequent as south-north migration has been in the past, and while 97 percent of the world's population does not move, migration is a global phenomenon that touches millions of lives, including many of those of haven't themselves migrated. (Stiglitz, Cortina & Ochoa-Reza, 2013, p. xv)

As the paragraph cited above tells us, women and girls account for half of international migrants. Even though popular opinion as well as most studies of migration still continue to perceive the immigrant as male, the reality is that the presence of females is central to the process (Cortina & Ochoa-Reza, 2013). With rapid mobility and more women migrating all over the world, migration "has taken on a gender dimension" (Chandra, 2011, p. 209). And yet, most published work on immigration does not focus on women *as women* or on the gendered processes that underlie the experience of migration. With very few exceptions, even when data about women and girls are presented, a gender analysis of the implications of these data tends to be absent.

The intent of this book is to provide a gender analysis of women's and girls' experiences of migration and to shed light on the often overlooked and understudied topic of psychological perspectives on women's experience of migration. The book fills a void in the psychological literature since it centers on explicitly feminist scholarship on immigrant women. We aim to present a gender analysis of migration, not just a "counting women" perspective. The purpose of the collection is not to present original empirical research on each and all topics but instead to incorporate the latest research literature and present theoretical reflections based on life data. The focus of the book is on applying this information and analysis to clinical practice and to the development of future research and theorizing in the fields of migration studies and feminist psychology.

A gender analysis does not simply examine women as subjects of scholarship but also explores ways in which gender is an organizing structure of power relations. "Conceptualizing gender as a process [rather than a variable], as one of several ways humans create and perpetuate social differences helps to deconstruct the myth of gender as a product of nature while underscoring its power dimension" (Mahler & Pessar, 2001, p. 441). Gender analysis interrogates the usual psychological focus on differences between women and men and questions power structures that determine and influence personal choices while hiding the ideological dimensions of those choices. A gender analysis recognizes that power structures influence the content, method and epistemology behind all psychological knowledge (Magnusson & Marecek, 2012). Therefore, while descriptions of findings concerning women and girls provide important information, we wanted to go beyond description. In other words, we chose chapters that do not simply examine data about girls and/or women but instead provide a feminist analysis in which gender is a central organizing axis of power, alongside other social structures such as age, class, race, ethnicity, nationality, and so on.

If we truly aim to understand the psychology of migration, then gender as a category of analysis, including gender relations and gendered power structures, cannot be ignored, trivialized or underestimated. To fully grasp the interplay of gender and the migratory process, it is important to understand that even though women migrate for a whole range of reasons, such as poverty, displacement, debt, and many other external constraints that they share with men, the impact of these problems is always gendered because of the impact of specific problems such as wage differentials in sender areas and in receiving areas (Kofman et al., 2000). Indeed, "all women face the constraints of sex segregation in employment" (Kofman et al., 2000, p. 25). In addition to these common economic needs and their gendered consequences,

> Some non-economic factors are especially important causes of migration for women. Social constraints facing women ... also influence sex selectivity patterns in migrations streams. Marital discord and physical violence, unhappy marriages and the impossibility of divorce [as well as other instances of sexual/gendered oppression] often influence women's decision to migrate". (Kofman et al., 2000, p. 21)

Furthermore, other triggers to migration, including political and individual reasons, are always already gendered, especially when considering, for example, issues of religion, education, war, sexuality, political instability or lack of natural resources, all as products of the state.

Women's agency, "viewed within the context of resisting oppression and exploitative structures ... is particularly vital for a gendered account of migration, because it is so often assumed that women simply 'follow' men and that their role in migration is reactive rather than proactive" (Kofman et al., 2000, p. 23). In other words, it is important to keep in mind not only that women's migration is triggered by financial opportunities dictated by men but also that immigrant women are also individual agents, often strategically seeking flight from patriarchal regimes (Kofman et al., 2000).

As such, at its roots, the phenomenon of international migration entails economic, political and gender issues. It involves both voluntary and involuntary migration. And although the usual understandings of international migration divide "true immigrants" as motivated by economic reasons and refugees a motivated by fleeing situations of danger, this division is not as clear when women's motivations for migration are concerned (Cole et al., 1992). States determine who is to be considered a refugee. This is not a self-assigned label. But the legal distinction between immigrants and refugees is colored by the political persuasion of governments. Therefore, these definitions tend to override the realities of the situation as experienced by the migrant herself. Is a woman fleeing marital violence in a country where the legal system does not protect her as a "voluntary" immigrant looking for an independent economic situation, or is she a refugee fleeing a situation of danger and thus an "involuntary" migrant? The lines are blurry because economics and physical danger interweave to create the motivation and the will to migrate. And the perception of governments and migrants may contradict each other in ways that produce material and institutional consequences.

Migration may offer women the opportunity to transgress established gender roles and challenge those traditional gender roles and power structures. Women's gender roles may shift because of their relocation to a country with more gender egalitarian norms or because of the necessity of their greater participation in the workforce (Espín, 1999). Women may view immigration as a chance to leave unsatisfying partnerships or improve their gender role status by increasing control over household decisions and building personal autonomy (Hondagneu-Sotelo, 1994, 1999, 2003). In other words, displacement may empower women to develop new selves. Many immigrant women perceive migration as providing "a space and permission to cross boundaries and transform their sexuality and sex roles," including embracing a non-heterosexual identity (Espín, 1999, 2006). The opportunities in work, education and routines of life provided by the migration are present in women's stories regardless of differences in the roles of women in their countries of birth.

Many women appear to benefit from the liberating effect of being "outsiders" in the new culture (Espín, 1999). But migration can also contribute to a situation where immigrant women embrace more traditional gender roles while embodying a "reimagined" home culture in their new cultural milieu (Mahalingam & Leu, 2005). Various social and cultural antecedents, such as experiences of previous trauma or challenges of maintaining transnational families, may place boundaries on women's ability or desire to transgress culturally prescribed gender roles. Traditional patriarchal contexts have always provided the opportunity to carve separate – if inferior – spaces for women. Migration carries with it the possibility of limiting women's private spheres of influence and their moral authority within traditional cultural contexts. The cultural transformations brought about by migration upset these spaces without yet giving women full access to equal power in the public sphere.

Indeed, migration can be both emancipatory and subjugating for women.

Women may benefit economically while their subordination as women may continue, although it may take different forms (Pessar, 2003). These tradeoffs and tensions create ambivalence and contradictions. The authors in this book explore these tradeoffs and tensions in different contexts and from different perspectives.

Feminist psychology not only focuses on individual women but also considers the impact of race, ethnicity, class, age, culture, sexuality and other sociohistorical/context factors. Further, we are interested in women's experiences in and of themselves, not simply as members of a family unit. Although families are an important factor, presenting women only as members of families – such as mothers, wives or daughters – misses the point that many women migrate alone and, in fact, encounter and negotiate the personal, economic, social and political issues raised by migration on their own. It is true that the work of immigrant women, married or not, is key to the survival of immigrant communities. It is also the case that many women who migrate alone do so in order to provide financial support for the families left behind in the home country. However, these economic contexts neither preclude nor suspend the personal reasons and circumstances that influence their decisions to migrate alone.

We are aware that migration is a global phenomenon. In assembling this book, our original wish was to include contributions that incorporate perspectives on migration concerning women and girls around the world. We cast a wide net, requesting contributions from many authors who work in the area of women's migration. But, as it happens with edited collections, we have been restricted by the topics submitted by authors. This collection has limitations. It is not as comprehensibly-global as we wished, nor does it cover as many perspectives from feminist psychology as we wanted. The focus is primarily on migration to North America.

By the same token, there are certain geographic areas of the world that are overrepresented in the book. Several chapters focus on women of Asian backgrounds and several focus on women from Latin America. This is not surprising when one considers that this geographical focus coincides with the reality of the largest immigrant groups in North America. However, this is another instance in which we have been limited by the responses we received from the array of prospective authors. Conversely, the positive side of emphasis on immigrants from some specific geographical areas and cultural groups is that it provides an in-depth look at the variability of experiences within those groups.

Similarly, we recognize that not all subject positions or groups are represented in this volume. For example, one chapter focuses on transgender women, while a specific exploration of lesbianism or heterosexuality is ironically absent. However, we believe that heterosexuality is very present in this volume, albeit implicitly, in discussions of motherhood and marriage. Indeed, "sexuality has almost always been relegated to and equated with the realms of heterosexual reproduction and family life" (Manalansan, 2006, p. 224). This does not mean that we do not value explicit interrogations of privileged identities or believe that they should continue to occupy the status of "normalcy" and thus escape critical examination. We look forward to more feminist scholarship of migration that unpacks whiteness, heterosexuality, masculinity, able-bodiedness and privileged class backgrounds (Dottolo & Stewart, 2013). The present legal situation concerning immigration for lesbians and gay men is in flux. The landscape is changing so rapidly that any chapter about lesbians and migration is bound to be obsolete by the time this book is published.

As one of our anonymous reviewers stated, this book fills several gaps in theory and research on the psychology of women, such as – but not limited to – intersectionality, violence against women and, obviously, migration itself as well as other aspects of the migratory process that are understood and experienced differently by women, men, girls and boys. We hope it pushes psychologists to think about the personal and emotional costs of the dislocation of space and language that permeates the contemporary global political/economic regime. We hope this edited collection, by bringing together multiple original essays can represent a whole that is more than the sum of its parts, which is of course what edited volumes do best. Together, the chapters give a sense of the variations in the migration experience along axes that complement and modify issues of gender in the specific migrant communities and in their relationships with the larger host society.

Some perspectives in this book focus on clinical interventions to help women with particular identities in particular circumstances: for example, Muslim women immigrants to the US or transgendered immigrant women. Other chapters are directed at improving the empirical foundations of theory and scholarship about intersectionality via the literature on women and migration.

While the topic of migration in general and in relation to gender has been studied extensively in recent years, this work has been done mostly from an anthropological, historical, sociological and/or public-policy perspective and much less from a psychological one. Women in certain immigrant groups (Latinas and Asians) which are also included in this book have indeed been studied from a psychological lens, including some book-length works (e.g. Kawahara & Espín, 2007, 2012). Some specific topics have been addressed (e.g. Espín, 1999) and several journal issues have focused on immigrant women. But to our knowledge, there is no single text that uses a gender lens to explore how migration experiences have implications for the psychology of identity and belonging, loss and longing, trauma and violence, sexuality and desire, memory, language, and intergenerational effects of migration through the lens of gender. We offer the present volume to break new ground in the psychological literature on gender and migration.

The extant psychological literature about women who migrate tends to pathologize their experiences and/or emphasize the needs of clinical populations. In other words, the focus tends to be on illness-based studies. "[P]sychologists have focused excessively on pathology. Much of the research emanating from the mental health field has searched for links between the migratory experience and expected negative psychological fallout resulting from that experience" (Suárez-Orozco & Qin, 2006, p. 185). By contrast, this volume emphasizes alternative themes, including those of survival, resilience and success without denying that migration is a highly stressful experience that influences immigrant and refugee women's well-being. Immigrant women may experience post-traumatic stress, mourning and grieving of multiple losses, acculturative stress, loneliness, loss of self-esteem, strain and fatigue from cognitive overload, uprootedness and perceptions that they are unable to function competently in the new culture. Yet, personality factors and psychological strengths – such as hope and optimism, cognitive flexibility and positive coping styles – also influence relocation experiences. No matter how strong the similarities created by the experience of migration, vast differences in personal experiences are present. Differences originate in the dissimilar countries of birth, ages at migration, historical period when the migration took place, circumstances surrounding it, pre-migration mental health issues and each individual's specific life experiences – as the chapters in this book ably demonstrate. Each immigrant woman carries a broader social history intertwined with her personal history.

Contrary to stereotypes, many women immigrants have high educational levels. Most chapter authors in this book are immigrants themselves or are US-born citizens who continue to be impacted by the migration experiences of their ancestors in ways that remain psychologically relevant for them. They are all successful scholars who demonstrate the psychological strengths they study and describe.

A note on the interplay of culture and gender roles

In compiling the chapters in this book, we never forgot that culture, with all its concrete and its intangible components, is at the core of who we all are. We could not be humans without a cultural context that supports families and individuals and also tells us who we are and what the world around us means. Culture creates our worldview and our identities as groups and as individuals. However, we are aware that "culture" is not a static entity. "[C]hange, conflict, and contradictions are resident *within* culture itself" (Phan, 2003, p. 13). And while cultural values and prejudices influence how immigrant women are represented in psychological science and cultural values are influenced by power, it is also true that the members of immigrant groups are cultural agents. We kept very present the reality that systemic forces of oppression and discrimination play a powerful role in shaping immigrant women's experiences, including their mental health and their access to psychological services and resources. But simultaneously, we are aware that groups that are transforming their way of life through a vast and deep process of acculturation focus on preserving "tradition" almost exclusively through the gender roles of women. Women's bodies become the site for struggles concerning disorienting cultural differences. Gender becomes the site to claim the power denied to immigrant men by the host society (Espín, 1999; 2006). This is not just a benign manifestation of interesting "cultural traditions." It may cost some women their lives.

Understanding the connections and disconnections between psychology and cultural issues seems to become more poignant when feminism and women's concerns are included in this mix. It brings us back to the question of who holds the power over meanings in psychology and inside a given culture (Magnusson & Marecek, 2012; Narayan, 1997; Phillips, 2010).

Pressures on immigrant women emerge from cultural values and attitudes present in the host society, which imposes its own burdens and desires on them through prejudices and racism. The role of power is extensive in the formation of cultural identity. In any process of acculturation, those who have more power can dictate terms of how and when "cultural" attitudes will be tolerated or rejected outright (Phan, 2003). Simultaneously, "returning women to their 'traditional roles' continues to be defined as central to preserving national identity and cultural pride" (Narayan, 1997, p. 20) by some immigrants. Those same values and behaviors may be perceived by their hosts as a demonstration of immigrants' "backwardness" and need for change.

We have to tease out the responsibility to respect cultures against an easy adjudication of power to represent the group to cultural *spokesmen* (Phillips, 2010). "In the name of equality between cultures, the guardians of conservative gender roles have been provided with public funds and offered an amplified voice on the public stage" (Phillips, 2010, p. 1–2). In an effort to be culturally sensitive, it is possible to forget that sexism does not deserve more respect when spoken in other languages. Sensitivity toward other cultures does not imply unquestioning acceptance of patriarchal definitions of cultural identities and behaviors.

Conversely, racism is not to be tolerated in the name of advancing women's rights. The balance between these two is a delicate but absolutely necessary one.

Needless to say, these attitudes have a serious impact in the provision of ethical and effective mental health services for and psychological treatment of women immigrants, as well as in the development of research questions that reflect material realities rather than preconceived ideas about immigrant cultures.

We should ask ourselves why the focus of "cultural tradition" is placed on women's roles and their sexuality rather than other aspects of culture and traditions. We need to reflect on the implications of respecting customs that target mainly women and have serious consequences for their bodily and emotional integrity. Deployment of "tradition" and "culture" to justify sexist, racist or homophobic behavior should never remain unproblematized (Espín, 2006). That is why, when trying to understand other cultures, we need to reflect on the implications of maintaining a customary system that targets mainly women. The question is how we preserve sensitivity and respect for others and their cultural differences while continuing to foster liberatory/emancipatory ideals and principles applicable to all oppressed groups, not just to some while excluding others.

It is not enough to be supported in efforts at liberation from ethnic/racial oppression if women's gender oppression as well as heterosexist oppression are not taken into consideration. The chapters in this book are an invitation to reflect on how we address our own racism, sexism, classism, heterosexism and other "isms" in a more careful way. This will demand more work to develop awareness about the interlocking nature of oppression and the continued weight of its many subtle forms – a daunting but necessary task. This requires paying attention and questioning the nature of power and the structures that perpetuate the exercise of that power –something, we might add, psychologists are not very good at doing (Espín, 2006).

These chapters challenge narrow conceptualizations of immigrant women as only oppressed and helpless. Many of the authors of these chapters are highly educated, immigrant women themselves. We believe that immigrant women from all social backgrounds are capable of strength, resilience and analyzing their situation.

It is very easy to assume that most poor, uneducated women immigrants do not understand the meaning and subtleties of women's oppression and to assume that those among them who do are probably just imitating or copying US, white, middle-class feminists. Within immigrant communities, it is common to dismiss any woman who expresses concern or opposition to male domination by labeling her as a "traitor" to her community and culture who has "sold herself" to "American" influence. The reality is that in all cultures (US and European cultures included) there will be resistance to women's transformations of their roles by those invested in maintaining the status quo. And, in all cultures women are capable of undertaking this transformative task. Those who have a stake in preventing the development of consciousness among women will not be pleased when those women take their lives in their own hands. Sometimes, apparent "cultural sensitivity" toward sexist practices is nothing but another variety of racism that, in fact, fosters a conservative politics which locks women into the past. Some of these practices perpetuate the "colonial gaze" and reinforce exclusionary practices used by those who have a vested interest in keeping women outside critical sites of power over their own lives (Espín, 2006).

However, it is important to recognize that some of the rigidity concerning the roles of women that we observe in immigrant communities is an attempt to protect and safeguard what remains of emotional stability; so much of it is lost with migration. That is why women immigrants themselves may join in this effort to preserve traditional gender roles. It is as if the immigrants' psychological sense of safety and their sense of self depended on a sharp contrast between two sets of cultural values conceived as rigidly different and unchangeable. The preservation of "old versions" of women's roles becomes central to this sharp contrast. For people who experience a deep lack of control over their daily lives, controlling women's sexuality and behavior becomes a symbolic demonstration of orderliness and continuity. Obviously, it is easier for immigrants to maintain control over their private world than over their public lives: work schedules, types of work and schooling, housing and the structures of daily life are controlled by the customs and demands of the new society. But in the privacy of their homes, they can seek to maintain the sense that they are still in control. This is why women frequently and actively join in adhering to "traditions" that, from the point of view of outsiders, appear to curtail their own freedoms and opportunities for self-fulfillment (Espín, 1999).

On the other hand, it is also essential to remember that expressions of male dominance among immigrants are nothing but the specific culture's version of the myth of male superiority that exits in most cultures, including mainstream US culture. And, although many immigrants of both sexes still subscribe to the traditional ideas of male superiority and its consequent forms of expression, many reject it outright, as it also happens in mainstream US culture.

In addition to acculturation patterns based on collective experiences of migration, individual factors play an important role in the process and speed of women's adaptation to their new environment. Successful acculturation may be influenced by women's age, economic resources, desire to adapt, the support of family and community and both cultural norms and cultural values, as well as women's ability to address psychological barriers, such as their previous traumatic experiences.

Immigrant women and girls develop their identity against the backdrop of these contradictions. Sometimes, these contradictions express themselves through emotional or other problems. More often than not, though, most immigrant women and girls remain unscarred and manage to emerge from the emotional struggle created by these incongruities and paradoxes. Despite prevalent misconceptions, women immigrants/refugees are "not only victims of structural forces or robots computing cost benefits of their moves" (Hondagneu-Sotelo, 1994, p. 6). They are also active agents in building their lives.

The authors of these chapters present us with multiple variations of these issues – such as the complex relationships between culture, identity and agency – in concrete situations and expressions.

A brief note on US immigration legislation and women

The significance of women's gender roles and sexuality in the process of migration becomes evident when observing the development of US immigration policy, which is largely based on misogynist ideas about gender and sexuality, particularly women's sexuality. Perhaps the simplest illustration is the importance placed by immigration legislation and enforcing authorities on issues such as prostitution, lesbianism, and pregnancy – all of which have been used as exclusion criteria for female immigrants at one point or another (Luibhéid, 2002).

Specifically, the first restrictive immigration law in the United States, the Page Law of 1875, focused on Asian immigration and placed particular emphasis on the enforcement of the law as it applied to Chinese *women*, assumed to be prostitutes unless proven otherwise (Peffer, 1986). Indeed, immigration legislation and enforcement has been a major site for the construction and regulation of immigrant women's identities. And through the control of immigrant women's sexuality, prevalent "moral" perspectives on all women in the US have been expressed. Yet, while there is well-developed scholarship about how official immigration policies aim to reproduce racial, ethnic, and class distinctions, there is minimal work on the role of US immigration policies in the creation of sexual and gender norms (Luibhéid, 2002).

A note on method

The chapters in this book provide different research, clinical and theoretical-narrative approaches. We wanted to present a variety of approaches while emphasizing the importance of theoretical analysis and reflexivity concerning each topic. We see reflexivity as both a practice and a methodology, rooted in feminist epistemology, that focuses on the researcher's/writer's reflecting upon the power relationships inherent in the research (or other scholarly endeavor) and making those power dynamics explicit. "Personal reflexivity involves reflecting on the ways in which one's values, experiences, interests, beliefs, political commitments, and social identities shape the research" (Magnusson & Marecek, 2012, p. 65). This is particularly important for authors who were studying issues close to their personal experiences. Therefore, we encouraged all authors to present a critical appraisal of their own practices: for example, why we frame issues or questions in particular ways; what drew us to our research questions; how we investigate and interpret; how our approaches lead us to particular kinds of knowledge and not others; and the ways in which our social identities (race, class, gender, sexuality, age, nationality, etc.) shape our relationships to the knowledge we produce.

This is why we want to acknowledge our own positionality in relation to the focus of this book. Oliva migrated alone for a variety of reasons to six different countries, over several decades. The particularities and commonalities of these migration experiences as a woman alone led to encounters and incidents that, most probably, would not have occurred had migration been undertaken together with family members – father, husband and brother – who might have been considered the "head of the household." These experiences provided contact with other women in similar situations. They created a clear awareness that looking at women immigrants only as members of families was missing a very important and large piece of women's experience of migration.

Andrea is the granddaughter of an Italian immigrant (her maternal grandfather) and was born and raised in an Italian American community in Syracuse, New York. The first in her family to attend graduate school, she lived in southern California and Michigan, both experienced as "different countries" in many ways. While intranational migration is clearly not the same, the scholarship on migration informed issues of longing, loss and memory, and they aided in the understanding of Americanized ethnic identities. As a feminist social identity theorist, it became clear that women's experiences of migration – either as immigrants themselves or as the daughters, granddaughters, and great-granddaughters of immigrant women – was underdeveloped in the scholarship.

In keeping with that focus, we wanted to include a selection of methodologies that contribute to enriching feminist psychologists' accounts of women's experiences of migration. Methodological differences among the chapters range from survey research to autobiography, from interviews to client narratives, from sociocultural analyses to structural power analyses in interpersonal relationships. This collection is thus methodologically rich, providing readers with refreshing, challenging, varied and innovative research and clinical exemplars.

While personal stories do not inherently make anyone's work feminist, we realize that stories of personal involvement may be likely to lend themselves to an analysis of reflexivity more than other approaches. The idea is to strike a balance between life experience and theoretical/critical understanding of whatever data is used to support the arguments of each chapter. Obviously, those chapters that are presenting data from empirical research studies, present reflexivity in a different perspective than those chapters presenting personal experiences. But we do believe that there is a variety oflegitimate and valid approaches and thus have tried to include them here.

We believe that the personal experience of the researcher is not only valid but essential in the development of studies that are meant to be contextual, reflexive and relevant. For Reinharz (1983), a relevant research project should provide "an opportunity for catharsis or self–discovery" (p. 176) for researchers and their subjects and a "research product likely to provide resources or answers to pressing problems in living" (p. 176). "The record of the researcher's feelings and ideas is also data" (p. 175) because "all knowledge is contingent on the situation under which it is formed" (p. 177) and because, for all researchers, "one's own race, class, religion and gender predispose us to consider some settings more interesting and important than others" (p. 179).

Similarly, the usefulness of life narratives for psychological research and the value of studying life histories and biography for the social sciences in general and psychology in particular has been demonstrated (e.g. Josselson, 1994; Josselson, Lieblich & McAdams, 2002; McAdams, 2006; Rosenwald & Ochberg,1992). "The data provided by life narratives produces a richness that could not be obtained through the use of other methods – generalizations drawn from interpretive research are not statements about individuals and populations, but rather are generalizations about cultural meanings and social processes" (Magnusson & Marecek, 2012, p. 68), and thus, they generate knowledge and contribute to theory development.

One of the reasons why the discipline of psychology has remained behind in the study of immigration and specifically issues of gender and migration (Suárez-Orozco & Qin, 2006) is its stubborn reliance on quantitative, positivist approaches. We could argue that this insistence and exclusive reliance on "hard" science is in itself a gendered statement about which approaches deserve to be valued, but that is a debate for another context. In any case, even though quantitative, positivist methods contribute significantly to knowledge in psychology, they alone cannot capture the subjective experience and agency of migrants, particularly women or the subtleties of gender ideologies. "Because [for the most part] psychologists insist firmly on a scientific methodology that emphasizes validity, reliability, and experimental designs, gender has been treated mainly as a dichotomous variable" (Donato, Gabaccia, Holdaway, Manalansan & Pessar, 2006, p. 17).

The slow advance of gender analysis in quantitative research has led to qualitative studies, narrative research and other approaches that lend themselves to conceptualize "gender as a central social category organizing the identities, social practices and institutions influencing migration" (Donato et al., 2006, p. 17). Because these approaches have been used by other disciplines, psychologists could benefit from interdisciplinary dialogue about methodology and epistemology in the study of migration. We hope the chapters in this book will contribute to that exchange. The authors demonstrate the richness of some alternative approaches to the development of knowledge beyond the usual quantitative approaches we have been all taught to use. This book takes the risk of "dancing through the minefields" (Kolodny, 1986; Marecek, 2003) of challenging traditional methods in the discipline of psychology and exploring new possibilities of inquiry.

In addition, traditional psychology has also been slow to understand and integrate the notion of intersectionality, where feminist scholars across disciplines have insisted upon its centrality for years. Intersectionality captures the realities of lived experience, that our identities cannot be separated from one another but instead mutually construct one another (Collins, 1990; Crenshaw, 1995; Dottolo & Stewart, 2008; Espín, 1997; Shields, 2008). Immigrant women cannot "choose" between being *either* an immigrant or a woman, because they occupy both identities simultaneously. Generally speaking, traditional psychology's insistence upon quantitative methods as most legitimate lends itself to this oversight and oversimplification, as complexities and multiple identities do not fit easily into data cells or ANOVAs. Indeed, intersectionality continues to be "a perspective in search of a method" (Shields, 2008, p. 306) in psychology and other "disciplines that prize methodological approaches that do not easily lend themselves to empirical study of intersectionality" (Shields, 2008, p. 301). Shields (2008) identifies other social sciences, including sociology and anthropology, as leaders of these scholarly paths and characterizes the need for psychology as a discipline to begin this journey as "urgent". However, "we do not aim to include all intersections of significance nor all of the innovative methodological strategies for investigation of intersectionality" (Shields, 2008, p. 307). The authors of these chapters have joined us in trying to develop "a useful roadmap toward those answers" (Shields, 2008, p. 307) as they pertain to the experiences of immigrant women.

Campbell and Wasco (2000) explain: "At an epistemological level, feminist social science legitimates women's lived experiences as sources of knowledge" (p. 775). They argue that one way in which this can practically be achieved is by expanding methodologies, including multiple ways of knowing, especially quantitative and qualitative methods, as part of a respected and valued set of tools. Since "the overarching goal of feminist research is to capture women's lived experience in a respectful manner that legitimates women's voices as sources of knowledge ... the process of research is of as much importance as the outcome" (Campbell & Wasco, 2000, p. 783). Indeed, the complexities in women's lives demand a multiplicity of research and clinical approaches (Olesen, 1994). We believe that the process is even more important when the focus is on the complex lives of immigrant women.

Yet, we are very well trained to doubt the validity of our personal experiences and life as sources of data. This book is thus designed to make visible and valuable the personal experiences that link authors and subjects of study in migration experiences. As Rosenwald and Ochberg (1992) tell us, "Personal stories are not merely a way of telling someone (or oneself) about one's life; they are the means by which identities may be fashioned" (p. 1). In order to provide an edited collection in which life stories are fully centered, we decided as editors to explicitly value multiple methodologies: feminist psychology, narrative research and interdisciplinarity – central tenets of the larger "story" of this volume.

Organization of the book

We have organized the chapters into four sections and provide a short introduction to each one of those sections in addition to this introduction and the Conclusion to the book. The sections present a progression that begin with general issues in immigrant women's lives in the new receiving/host context, as influenced by origin/home culture (Section 1).

This is followed by a section on economic issues (Section 2), and one on issues of violence (Section 3). The last section (Section 4) focuses on intergenerational effects migration on women who are descendants of immigrants.

Part 1: Place, race, memory and migration

Espín situates memory in the psychology of immigrant women's lives, exploring how the reconstructions of events over time, place and space shape and inform identity and experience. She explores how gender, language, race and class shape immigrant themes of loss, longing, separation and home, and she discusses the implications for therapy of understanding the interplay of memory and life events.

She contends that for all immigrants, issues of loss, grief and mourning are of primary psychological importance. And so also is the significance of place – the lost home and the one that has been gained through migration. Memory of what was and what could have been are frequently very present in immigrant women's lives. Exploring how the reconstruction of events over time, as well as place and space, shape and inform the identity and experience of immigrant women is an important consideration in understanding their psychology.

Suyemoto and Donovan explore how intersections of power, privilege and oppression related to race, gender and nativity influence in-group relations among women of color. Specifically, the experiences of black and Asian immigrant and US-born women as they negotiate the dominance of race as a social status in the United States, the influence of imposed meanings of gendered racialization and their relation to nativity, and ways these imposed meanings affect relations between immigrant and US-born black and Asian women within the same racialized group are addressed.

The development of identity and adaptation to a new life is further problematized for women and girls who come from non–European backgrounds. They are confronted with the realities of racial politics in the United States in addition to other acculturation issues. Because they are "racialized" individuals immersed in racial politics, their adaptation to the host society is additionally burdensome for them. Racialized immigrant women are assumed to be members of "ethnic/racial" minorities which poses its own conflicts and dilemmas as Suyemoto and Donovan discuss.

Together, these two chapters set the tone of the book and present the main issues and contexts that the book's other chapters explore.

Part 2: Work, social class and "traditional" gender roles

This section features chapters that highlight immigrant women's economic realities while negotiating gendered expectations in the host country – especially connected with themes of work, employment and social class – and their impact on women immigrants' experiences of mothering. Grounded in a feminist family therapy perspective, Ma and Shea explore the literature on East Asian immigrant women's experiences with employment and mothering. Gender- and culture-based parenting ideologies and the intersection of acculturation, ethnicity and social class are highlighted. Suggestions for clinical interventions with this population are provided.

Balan and Mahalingam present an empirical investigation of the psychological contours of the model minority stereotype present in and about Asian immigrant communities using a combination of an online survey and semi-structured interviews. Intersecting immigrant cultural expectations (e.g. the model minority stereotype) and gendered processes (e.g. motherhood and employment) are addressed, especially ways in which they shape immigrant working mothers ideals about being "good Asian moms".

Through detailed conversations with Afghan women, Ahmed-Ghosh discusses their experiences of migration to the US. What makes the position of Afghans as refugees distinctive in the US is that they are fleeing their home country to become refugees in the very country that is at so-called war with theirs. Mothering is a complicated process for Afghan refugee women. The responsibility of cultural transmission rests mostly on their shoulders, despite their children's progressive acculturation away from those traditions and Muslim religious norms. This chapter discusses how gender and class differences based on the wave of migration from Afghanistan, the present class situation in the US, and the US-Afghan conflict, affect these women experiences as mothers and as Muslims.

Part 3: Violence, resistance and resilience

This section addresses the often silenced role of violence in immigrant women's experiences, including sexual, domestic and political violence.

Kallivayalil uses a feminist and trauma-informed psychological perspective to investigate the complex challenges faced by women who are victims of politically motivated violence and who are seeking political asylum. Clinical case material is utilized to highlight ways of working with this population in a culturally informed manner. Tummala-Narra, Satiani and Patel provide a gender analysis of aspects of South Asian culture, including religion and spirituality, sexuality and family structures and dynamics. Using a feminist psychodynamic lens, this chapter aims to extend existing clinical understanding of social and contextual issues that contribute to the problem of sexual violence, the recovery process and empowerment within South Asian communities.

Quintero, Cerezo, Morales and Gebhardt discuss a community-based qualitative research project that investigated the lived experiences and consequent mental health needs of transgender women who migrated to the US from Latin America. Experiences of violence, addiction and mutual support described by the women interviewed, particularly Erika (the case study that is the focus of this chapter), are central to their analysis of the situation of immigrant transgender Latinas. Recommendations for professionals' development to better understand and intervene in the lives of transgender immigrant Latinas are included.

Serrata, Macías, Rosales, Rodríguez and Perilla present the discoveries of a study of a group of Latina immigrant survivors of domestic violence who completed a leadership development program (the *Líderes* program) in the Southeastern United States. Phenomenological data analysis is used to identify themes of the importance of acknowledging histories of violence, the importance a supportive environment from which the program was delivered, personal and observed discovery of natural leadership abilities and a collective commitment to creating community change. The program described in this chapter is an illustration of the role of resistance and resilience in the lives of immigrant women who are survivors of violence in its many forms.

Part 4: Intergenerational impact of migration

This section explores the effects of migration on the daughters and granddaughters of immigrant women, including themes of acculturation, identity and the psychological residues of gendered journeys. Stutman and Brady-Amoon focus on the US citizen daughters of undocumented Ecuadorian immigrants to the US at a moment in time when one or both parents has emerged from hiding in order to attempt to gain legal residential status in the United States. The authors explore the lived experiences and stressors of 39 Ecuadorian American girls as told through the narratives that accompany their House-Tree-Person (HTP) drawings, and conclude with recommendations for practice and policy. This chapter illustrates one of the issues in the intergenerational transmission of immigration stresses.

Mattar and Pestana address some aspects of the gendered immigration and acculturation gap in the literature by discussing the need to understand migration narratives across generations and the impact of multiple contextual factors on acculturation processes. Using a case study method, the chapter presents the first-person immigration narratives of two generations of women impacted by voluntary transnational serial migration.

Dion considers the challenges that gendered expectations can create for daughters from immigrant families as they construct a sense of their own identity. Based on interviews with university undergraduate women born in Canada whose parents were immigrants to that country from several different societies, Dion provides an approach to self-definition for understanding gender and cultural identity.

Dottolo and Dottolo focus on how granddaughters of immigrants construct and maintain their ethnic identity – in this case, being Italian. Since language cannot serve as a symbol for these women, because speaking Italian was often forbidden in their homes or spoken only between adults in covert communications, they must cling to other symbols of Italianness in order to preserve their sense of gendered ethnic identities. The chapter argues that one such symbol is food, wherein participants manipulate recipes and food to navigate and negotiate being both Italian and American, femininity, social class, and honoring both a lost and imagined cultural experience.

Concluding remarks and reflections

We have delighted in compiling this volume and value its significance in addressing and redressing the absences and silences of women's experiences in feminist psychological perspectives of migration. The title of this collection reflects many journeys – the physical crossing of state borders; the transgression of cultures; and their roles and norms around gender, race, class, sexuality, age, ability, religion and citizenship. It also represents the journeys of the authors, poring through the data of their participants, their interviewees, their clients and their own lives. As editors, we have had the privilege of accompanying them, and each other, on their adventures in contributing to this groundbreaking work. Since we have highlighted reflexivity as an important theme of these chapters, the title of this book also reflects our own journeys in our relationship with each other – student, teacher, mentor, trusted adviser, friend, author, editor, traveling companion and intellectual playmate.

While we are proud to be part of this contribution to the scholarship on women's experiences of migration, we also hope that it generates more questions than answers, as all good theory and research do. For example, what *other* ways can we understand women's experiences of migration? Who is not being seen, heard or represented? What issues and topics need further investigation? What are the consequences of our omissions? How might the understandings of migration presented in this volume change over time, as communities become ever more connected? Who will shape the landscape as its contours continue to shift and its boundaries and borders are redefined? How will psychology respond to and engage with scholarship on migration that prioritizes intersectionality, interdisciplinarity and reflexivity, to name a few?

Whatever the answers to these questions, we hope that feminist psychology continues to be part of the conversation.

References

Campbell, R. & Wasco, S.M. (2000). Feminist approaches to social science: Epistemological and methodological tenets. *American Journal of Community Psychology, 28*(6), pp. 773–791.

Chandra, P.S. (2011). Mental health issues related to migration in women. In D. Bhugra & S.

Gupta (Eds.), *Migration and mental health* (pp. 209–219). New York: Cambridge University Press.

Cole, E., Espín, O.M. & Rothblum, E.D. (Eds.) (1992). *Refugee women and their mental health: Shattered societies, shattered lives.* New York: Haworth Press.

Collins, P.H. (1990). Black feminist thought: Knowledge, consciousness and the politics of empowerment. London: Harper Collins.

Cortina, J. & Ochoa-Reza, E. (2013). Women, children, and migration. In J. Cortina & E.Ochoa-Reza (Eds.), *New perspectives on international migration and development* (pp. 139–163). New York: Columbia University Press.

Crenshaw, K. (2005). Intersectionality and identity politics: Learning from violence against women of color. In W.K. Kolmar & F. Bartkowski (Eds.), *Feminist theory: A reader* (pp. 533–542). Boston: McGraw Hill.

Donato, K.M., Gabaccia, D., Holdaway, J., Malanasan, M. & Pesar, P. (2006). A glass half full?Gender in migration studies. *International Migration Review, 40*(1), pp. 3–26.

Dottolo, A.L. & Stewart, A.J. (2013). "I never think about my race": Psychological features of white racial identities. *Qualitative Research in Psychology, 10*, pp. 102–117. doi:10.1080/14780887.2011.586449.

Dottolo, A.L. & Stewart, A.J. (2008). "Don't ever forget now, you're a Black man in America": Intersections of race, class and gender in encounters with the police. *Sex Roles, 59*, pp. 350–364.

Espín, O.M. (2006). Gender, sexuality, language and migration. In R. Mahalingam (Ed.),*Cultural psychology of immigrants* (pp. 241–258). Mahwah, NJ: Lawrence Erlbaum.

Espín, O.M. (1999). *Women crossing boundaries: A psychology of immigration and thetransformations of sexuality*. New York: Routledge.

Espín, O.M. (1997). *Latina realities: Essays on healing, migration and sexuality*. Boulder, CO: Westview.

Hondagneu-Sotelo, P. (1994). *Gendered transitions: Mexican experiences of immigration*. Berkeley, CA: University of California Press.

Hondagneu-Sotelo, P. (1999). Gender and contemporary U.S. immigration. *American Behavioral Scientist, 42*, pp. 565–576.

Hondagneu-Sotelo, P. (Ed.) (2003). *Gender and U.S. migration*. Berkeley, CA: University ofCalifornia Press.

Josselson, R.E. (1994). Identity and relatedness in the life cycle. In H. Bosma, T. Graafma, H.Grotevant & D. De Levita (Eds.), *Identity and development: An interdisciplinary approach* (pp. 81–102). London: Sage Publications.

Josselson, R.E., Lieblich, A. & McAdams, D.P. (2002). *Up close and personal: The teachingand learning of narrative research*. Washington, DC: American Psychological Association.

Kawahara, D. & Espín, O.M. (Eds.) (2007). *Feminist reflections on growth and transformations: Asian-American women in therapy*. New York: Haworth Press.

Kawahara, D. & Espín, O.M. (Eds.) (2012). *Feminist therapy with Latina women: Personal andsocial voices*. New York: Routledge.

Kofman, E., Phizacklea, A., Raghuram, P. & Sales, R. (2000). *Gender and internationalmigration in Europe: Employment, welfare and politics*. New York: Routledge.

Kolodny, A. (1986). Dancing through the minefields: Some observations on the theory, practice,and politics of a feminist literary criticism. In H. Adams & L. Searle (Eds.), *Critical theory since 1965* (pp. 499–512). Tallahassee: Florida State UP.

Luibhéid, E. (2002). *Entry denied: Controlling sexuality at the border*. Minneapolis, MN: University of Minnesota Press.

Magnusson, E. & Marecek, J. (2012). *Gender and culture in psychology: Theories and practices*.New York: Cambridge University Press.

Mahalingam, R. & Leu, J. (2005) Culture, essentialism, immigration and representations of gender. *Theory & Psychology, 15*(6), pp. 841–862.

Mahler, S. & Pessar, P. (2001). Gendered geographies of power: Analyzing gender across transnational spaces. *Identities, 7*(4), pp. 441–459.

Manalansan, M.F. (2006). Queer intersections: Sexuality and gender in migration studies. *International Migration Review, 40*(1), pp. 224–249.

Marecek, J. (2003). Dancing through minefields: Toward a qualitative stance in psychology. In P. Camic, J. Rhodes & Yardley (Eds.), *Qualitative research in psychology: Expanding perspectives in methodology and design* (pp. 46–69). Washington, DC: American Psychological Association.

McAdams, D.P. (2006). *The redemptive self: Stories Americans live by*. New York: Oxford University Press.

Narayan, U. (1997). *Dislocating cultures: Identities, traditions, and third world feminism*. New York: Routledge.

Olesen, V. (1994). Feminisms and models of qualitative research. In N.K. Denzin & Y.S. Lincoln (Eds.) *Handbook of qualitative research* (pp. 158–174). Thousand Oaks, CA: Sage Publications.

Peffer, G.A. (1986) Forbidden families: Emigration experiences of Chinese women under the Page Law, 1875–1882. *Journal of American Ethnic History, 6*(1), pp. 28–46.
Pessar, P.R. (2003). Engendering migration studies. In P. Hondagneu-Sotelo (Ed.), *Gender and U.S. migration* (pp. 20–42). Berkeley, CA: University of California Press.
Phan, P.C. (2003). *Christianity with an Asian face: Asian American theology in the making*. New York: Orbis Books.
Phillips, A. (2010). *Gender and culture*. Cambridge: Cambridge University Press.
Reinharz, S. (1983). Experiential analysis: A contribution to feminist research. In G. Bowles and R.D. Klein (Eds.), *Theories of women's studies* (pp. 174–182). London: Routledge & Kegan Paul.
Rosenwald, G.C. & Ochberg, R.L. (Eds.) (1992). *Storied lives: The cultural politics of self-understanding*. New Haven, CT: Yale University Press.
Shields, S.A. (2008). Gender: An intersectional perspective. *Sex Roles, 59*, pp. 301–311.
Stiglitz, J.E., Cortina, J. & Ochoa-Reza, E. (Eds.) (2013). Introduction. In J. Cortina & E.Ochoa-Reza, *New perspectives on international migration and development* (xv–xx). New York: Columbia University Press.
Suárez-Orozco, C. & Qin, D.B. (2006). Gendered perspectives in Psychology: Immigrant originyouth. *International Migration Review, 40*(1), pp. 165–198.

Part I

Place, Race, Memory and Migration

Introduction to Part I

Andrea L. Dottolo and Oliva M. Espín

The chapters in this first section serve to provide a topical, theoretical and epistemological frame for this collection. The chapters by Espín as well as Suyemoto and Donovan serve as a guidepost to this collection in two formative ways.

Espín situates memory in the psychology of immigrant women's lives, exploring how the reconstructions of events over time, place and space shape and inform identity and experience. She explores how gender, language, race and class shape immigrant themes of loss, longing, separation and home, including a discussion of the implications for therapy of understanding the interplay of memory and life events.

Suyemoto and Donovan explore how intersections of power, privilege and oppression related to race, gender and nativity influence in-group relations among women of color. The experience of immigrant and US-born women as they negotiate the dominance of race as a social status in the United States, the influence of imposed meanings of gendered racialization and their relation to nativity and ways these imposed meanings affect relations between immigrant and US-born women are addressed. This chapter presents a theoretical understanding of the intricacies of relationships among racialized women both immigrants and US-born black and Asian, based on the firsthand experience of the authors.

The title of Espín's chapter includes the phrase "geography of memory", and taken together, these two contributions could be entitled, "the geography of identity" – referring to the identity processes of immigrant women that are examined across the chapters, as well as the "identity" of the book itself. These chapters extend the notion of intersectionality to include not only the ways in which race, class, gender and sexuality mutually construct and define each other, but also take account of nativity, language, history, memory and place, to name a few, which are also reflected in the sections and chapters to follow. Espín discusses issues that are common among all immigrant women, such as memory, place and loss, and Suyemoto and Donovan situate their experiences within institutional structures and social processes of power and oppression.

These two contributions map the terrain of the volume, marking the landscape and surveying the territory, ground and scope of the book. They do so by pointing out themes, struggles and methodologies of this collection, which is directed at improving our understanding of the lives of immigrant women. By surveying the notions of place, race and memory in migration, they serve as an umbrella for contributions about work, social class, "tradition," violence, resistance, resilience and the intergenerational impact of migration. Both Espín as well as Suyemoto and Donovan outline many of the central psychological features of women's experiences of migration, including memory, grief and the construction of the "other". Both chapters also highlight the complexities and contradictions in the psychologies of immigrant women.

One especially important way in which both Espín as well as Suyemoto and Donovan situate a frame for this volume is through their explicit analysis of the power of narrative. These authors not only reveal themselves, exposing their vulnerability in their stories and in the mere act of telling them, but then acknowledge and interpret the baring and uncovering as part of their method and as a source of strength. Suyemoto and Donovan use reflexivity as a tool of analysis in understanding their own stories and their relationship with each other. Espín extends traditional psychological notions of talk therapy to encompass multiple sophisticated layers of healing through narration in the lives of immigrant women. Each of the subsequent chapters in this collection tells an important story of women's experiences of migration using multiple methods and perspectives.

These chapters serve as a foundation for the book in ways in which the authors ground their work to the central feminist tenet that "the personal is political". Espín discusses how her memory and very sense of self are products of sociohistorical events, connecting her life story to a larger chronicle of gendered migration. Suyemoto and Donovan also use personal experiences and an explicit analysis of "the personal as political" as a point of departure in reflecting upon relationships between immigrant women of color and US-born women of color. Memory, identity, grief, longing, feelings of belonging and otherness are all products of larger institutional and political structures, shaping and informing each other.

Espín as well as Suyemoto and Donovan also model the ways that the subsequent chapters explore questions about the ways in which migration influences how individual women and women as a whole understand themselves in relation to their immediate social structures.

Social identity theorists in psychology have been late to consider migration, place and space, or even power and privilege, in the development of self. Everyone must navigate a social terrain of imposed categories and attributions in the process of identity negotiation. Dottolo and Stewart (2013; 2008) have contributed to the emerging discourse on the ways in which institutional structures shape social identities. However, those with marginalized identities often have an additional challenge of learning to embrace an identity that is culturally undervalued or derided (Espín, 1997).

Theologian Fernando Segovia (2001) describes immigration to the US as "a mangled project of exalted principles and despicable behavior" (p. 267). However, psychology has yet to incorporate into its scholarship the historical and social complexities and contradictions that have influenced US immigration policies and how these policies have shaped individual lives. The authors in this collection demonstrate that both "exalted principles" and "despicable behavior" are present in the experiences of women immigrants. Generally speaking, our undergraduate students are still being taught about one such "exalted principle," that the United States is a "melting pot" and that "assimilation" is and should be the goal of all immigrants, while instances of "despicable behavior" are muted, marginalized or ignored. Both "exalted principles" and "despicable behavior" translate in the personal as political.

The point that the personal is political also ties to another theme of these two chapters that help to set the stage for the book – the social construction multiple identities and realities. Immigrant women in the US experience multiple sources of social oppression. These forces include, but are not limited to, xenophobia, racism, sexism and discrimination based on poverty and employment. Espín ponders the multiple selves that may have resulted from staying in Cuba. She explores the validity, accuracy and emotional valence of memories and how they shape who we think we are. Donovan describes the social construction of racial formations, explaining that back in Guyana, she did not conceive of herself in racial terms, and only upon arrival to the US did she quickly learn that "race mattered here". Suyemoto articulates the pain and tensions around the "border patrolling" of who gets to be a "real" Asian and who is not.

The two chapters in this section foreshadow the purpose, significance and intention of the volume as a whole by claiming speech as a political act. Telling the stories of immigrant women, as women in and of themselves; subjects, and not objects of knowledge; carving out a space in the discourses of migration; agentic and defiant, is by definition an act of power. As bell hooks (2008/1993) explains,

Moving from silence into speech is from the oppressed, the colonized, the exploited, and

> those who stand and struggle side by side a gesture of defiance that heals, that makes new life and new growth possible. It is that act of speech of "talking back," that is no mere gesture of empty words, that is expression of our movement from object to subject – the liberated voice. (p. 18)

Espín as well as Suyemoto and Donovan serve as models, as templates, and together as an introduction to a collection that we believe does just that.

References

Dottolo, A.L. & Stewart, A.J. (2013). "I never think about my race": Psychological features of white racial identities. *Qualitative Research in Psychology, 10*, pp. 102–117. doi:10.1080/14780887.2011.586449.

Dottolo, A.L. & Stewart, A.J. (2008). "Don't ever forget now, you're a Black man in America": Intersections of race, class and gender in encounters with the police. *Sex Roles, 59*, pp. 350–364.

Espín, O.M. (1997). *Latina realities: Essays on healing, migration, and sexuality.* Boulder: Westview Press.

hooks, b. (2008/1993). Talking back. In A. Kesselman, L.D. McNair & N. Schniedewind (Eds.), *Women, images and realities: A multicultural anthology* (pp. 15–18). New York: McGraw Hill.

Segovia, F. (2001). Melting and dreaming in America: Visions and re-visions. In E.S. Fernandez & F.F. Segovia (Eds.), *A dream unfinished: Theological reflections on American from the margins* (pp. 228–267). New York: Orbis Books.

1
A Geography of Memory: A Psychology of Place

Oliva M. Espín

My life has been marked by the experience of migration; I have lived most of my life removed from my country of birth. As I wrote several years ago,

> I possess the vague certainty that I could have been another person were it not for the particular circumstances that migration brought into my life. I do not know and will never know the person I could have been had I not left my country. The only me I know is the one that incorporates the consequences of migration. Even though my life has been very rich in experiences, and I have never felt particularly deprived, I know that whatever I have succeeded in creating and living has been developed at the expense of some significant losses. Of these losses I am only vaguely aware. Far more clear are the undeniable opportunities, achievements, successes, and fulfillments brought about by migration. For the person who has migrated, identity issues are further complicated by their polyvalent circumstances. One is aware that both life's losses and failures and life's possibilities and triumphs are magnified and distorted by the lens of the migration experience. Migration for me, as for most immigrants, has given a dual and contradictory legacy. It provided safety and success, yet it also brought losses and silence about them. Mention of them is easily confused with self-pity or even ungratefulness to the new country. (Espín, 1999, p. 1)

In 1961 I left Cuba, my country of birth, and started life as an immigrant in Spain, Panama and Costa Rica. Later circumstances brought me to Belgium and to Canada. Finally, I came to reside in the United States. Although I have been a citizen of this country for several decades, my memories of childhood and adolescence in Cuba and the memories of early adulthood in other countries continue to be at the core of who I am.

Historical events and individual lives intersect to create autobiographical memory. The historical events I have witnessed or lived through in different parts of the world have served to frame my life in reality, to give it a sense of groundedness despite the many breaks and dislocations that have characterized it. Conflicts that occurred in places where I was living at the time – such as the Cuban Revolution, including the Bay of Pigs Invasion or the incidents at the Panama Canal Zone in the early 1960s – serve to anchor my memories. With aging has come a clearer sense of the importance of the events I have witnessed and of how entangled they have been with the fabric of my life, as well as the need to understand at a deeper level what role they have played in the weaving of that fabric.

In this chapter I discuss the role of memories in the formation, preservation and integration of self and identity, particularly for immigrants, as well as the role of language in the encoding or decoding of those memories. The importance of memories of place and the place of memories – what I call a "geography of memory" – are central to my discussion.

Memory and identity

Some psychologists who study autobiographical memory (e.g. Rubin, 1986) believe that we *are* our memories, that we are what we remember, real or imagined. Although autobiographical memory is important for everyone, for immigrants or those who have lived in many places throughout their lives, memory provides the only sense of continuity. Places and people change; the only way of knowing who you are – that you are – is to remember. Memory functions to construct a sense of continuity in our lives. The thread created by memories binds multiple experiences into a sense of personal identity and subjectivity – a sense of continuity to feel whole, which can be summarized as "who I am" (Taylor, 2010).

When powerful dislocations occur in life, as in migration, memory may be the only tool to recover a sense of self. Memory is the only witness to immigrant lives. It is the only way to *re-member* all the scattered pieces of life. As Chilean author Isabel Allende explains, "[t]hose of us who have moved on many times... lack roots and corroboration of who we are, we must put our trust in memory to give continuity to our lives" (2003, p. 79).

However, psychologists know that memory is precarious, fragile and changeable. Psychological and physiological studies demonstrate that memory is not always reliable, that autobiographical memory can be distorted, that memory is affected by many factors (e.g. Kandel, 2006; Schacter, 1996; 2001), including the languages in which memories are encoded and decoded (e.g. Javier, Barroso & Muñoz, 1993; Schrauf, 2003; Schrauf, Pavlenko & Dewale, 2003). But memory betrays, it mixes events and it leaves holes precisely about the moments you need to remember most. "Memory is always cloudy, we can't trust it" (Allende, 2003, p. 79).

Absence of places and people and even familiar smells create a discursive obstacle in the creation of identity that is unavoidably filled with "remembered" imaginings. Memory is not just a residue or a fixed reproduction of events but also a dynamic, conflictual and constructive process, deeply intertwined with values, communications, language and feelings, to name a few. Emotion both intensifies and obliterates the memory of events.

On the positive side, as cognitive psychologist David Pillemer (2000) explains, "although examples of memory distortions exist... those memories that do persist into later life are likely to generally be truthful" (p. 56). Memories of peripheral details tend to be more susceptible to distortion than the memory for central themes, which tends to be very reliable. Pillemer tell us "that we can largely trust our vivid memories of emotional events (p. 56) [because] emotional memories, although not infallible, are often broadly accurate" (p. 55). "Memories of apparent trivial details can [also] fulfill important psychological functions" (p. 62). Critical insight and engagement also have a transformative effect on memory and narratives.

I know that my individual memory is simultaneously my own process, connected to the nerve cells in my brain *and* the product of relationships and social contexts that have shaped it to determine what is important to remember. I also know that, to some extent, all childhood memories are imaginary. Memories of childhood idealize and distort; they may be full of resentment or rosy fantasies. The inevitable lacunae get filled up with those imaginations. Although adult memories can be more stable, they are also subject to multiple factors that influence what and how we remember "facts".

And yet, despite all its limitations, all I really have is memory.

Memory, culture and society

In addition, memories and self-understanding are "always shaped by culture. The tales we tell each other (and ourselves) about who we are...are individual variations on the narrative templates our culture deems intelligible" (Ochberg, 1992, p. 214). What is remembered has a great deal to do with what is culturally and theoretically acceptable at a given point in history. As in all identity development processes, individual desire and societal possibilities both push the limits and constrain the boundaries of the lived story and of the memories that sustain that story. When societal transformations occur, when people find themselves in a different cultural context, the acceptable accounts of life stories are also transformed.

Stories lived and stories told by people, acceptable behaviors and acceptable accounts of behavior are regulated by society (Rosenwald, 1992). The social construction of past events has an enormous impact on the structuring of present day social and political realities as well as on the meaning of our individual lives (Rosenwald & Ochberg, 1992). "Insufficient attention to the accumulated meanings given by society and culture and their inescapable impact on our understanding of our world and ourselves" (Taylor, 2010, p. 26) interfere with understanding individual lives.

Collective narratives of "who we are" interact with our individual histories to create our sense of identity and our way of interpreting our surroundings. Collective narratives are schemes by which we make meaning of "our history". They do not appear out of nowhere but are created from the way groups reconstruct their history and identity. In the first part of the twentieth century, French sociologist Maurice Halbwachs (1941/1992; 1950/1980) devoted considerable attention to the social construction of memory. According to Halbwachs, individual human memory is possible only within a collective understanding. Our ideas about the present are based on collective reconstructions of the past. According to Halbwachs, individuals organize and understand events within a social context, and therefore, they remember those events in the light of the collective understanding. Halbwachs believed that individuals' understanding of themselves and of historical events is linked to their social group's consciousness (Halbwachs, 1941/1992; 1950/1980). The group constructs the memory, and individuals do the remembering within those parameters. Collective memories are always selective; therefore, different groups may have different memories of the same event, which in turn affect modes of behavior for both individuals and groups. To illustrate his theory Halbwachs (1941/1992) used examples of how memories of the past in wealthy families in France diverge considerably from those of other French citizens. As an additional illustration he showed how changes in the memories of pilgrims to the Holy Land through the centuries depended on their historical moment.

Other authors have affirmed that collective historical memory is a powerful force that shapes who we are as individuals (e.g. Salomon, 2004). As Salomon states, citing Ignatieff (1996), "...the collective memory of the past isn't past at all" (Salomon, 2004, p. 275). It plays an enormous role in how we remember the events of our lives and what we remember about them, which is to say that collective memories contribute to distortions in the individual memories about "what really happened".

Several reference groups compete for the loyalties of women immigrants. The collective memories on which they base their own are not only those of the history of their country of birth but also those of the new country and of the immigration process itself. Each migratory wave arrives with its own particular socio-historical location and is received by a host country whose socio-historical processes and consciousness are also being transformed. Grievances and guilt as well as hope and anticipation that are experienced by individual immigrants are determined by collective interpretations of the "essence" of both nations and of the migration itself. They are expressed in individual behavior and experienced as personal by the individual immigrant.

Immigrant communities frequently "invent traditions" (Hobsbawm & Ranger, 1992) as a way of coping with the stress of the new environment by holding on to some aspects of the real or imagined past. These invented traditions may include distortions of "how things were" back in the country of origin or may consist on preserving a past that is no longer viable in the home country. This is an effort to preserve values, beliefs and norms that safeguard a social identity and the authority of some members of the community. For women, issues of gender become entangled with this need to preserve the past at all costs in the midst of rapid change. Their loyalties to group and family may conflict with their developing identities in the new context. Collective understanding of what it means to be a woman in a particular culture may conflict with expectations in the new country. They frequently become a source of conflict in immigrant families, as therapists and researchers have observed (e.g. Espín, 1999).

Memory through the generations

Migration also produces psychological effects for the immigrants' descendants beyond the obvious fact that the younger generations are born and/or grow into adulthood in the new country rather than the country of their ancestors. Migration influences the nature of parenting and other relationships. Parental behaviors may be influenced by the expectations of the new surroundings concerning family roles. Younger children of immigrants may experience different parental behaviors than what their older siblings experienced because parents may respond to the demands of the new environment.

The intergenerational transmission of trauma combines with the psychological transformations prompted by the migration to produce multiple effects. Post-traumatic stress, mourning and grieving, acculturative stress and other phenomena that are common among immigrants and refugees impact the lives of their children, even though those children have not experienced the process of migration in their own lives.

Descendants of immigrants do not have memories of any place other than the one where they live. And yet, their parents' memories of another place that the children may have never seen continue to haunt them in their present life and location and continue to influence their own sense of identity. Children of immigrants "received the emotional traces of parents' experiences...but not memories" (p. 114). Memories are not genetically transmitted. Therefore, "whereas adults who live through violence and atrocity can understand what happened to them as actuality – no matter how awful its terms – the generation after receives its first knowledge of the terrible events with only childish instruments of perception, and as a kind of fable" (Hartmann, Hoffman, Mendelsohn & Miller, 2011, p. 114). "[M]emory is individual, you can't remember things that didn't happen to you" (p. 115). But "when you grow up in an immigrant family you are always hearing about the country of origin. So [if you ever go there] it feels like going back" (p. 112).

Paradoxically, the psychological effects of migration on children of immigrants are derived not only from losses and traumatic experiences of their parents but also from the successes immigrants may achieve as a consequence of the migration. New loyalties in the host country may be experienced by both immigrant parents and their children as betrayal of loved ones and the homeland. As Boszormenyi-Nagy and Spark (1973) put it, "invisible loyalties" can be emotionally paralyzing and can induce compulsive behavior in individuals and families who may feel guilty if they "forget" what happened in the old country. "Invisible loyalties" may interfere with adaptation to the new country. Indeed, the immigrant's "sense of their own identity may be profoundly disturbed, if they feel that adaptation requires them to betray their earlier attachments" (Marris, 1974, p. 82).

The "geography of memory"

When violent or tragic events have occurred in a particular place, the memories of that place become "contaminated" by those events. According to cultural geographer Kenneth Foote (1993), the physical space impacted by those events requires some form of reaction to the specific place that can range from sanctification to obliteration of a given place. Elizabeth Jelin (2003) and other South American authors have studied memories as they pertain to events and places impacted by the dictatorships in the 1970s and 1980s in the countries of the Southern Cone. Jelin and many others have used memory studies during the last twenty years to explore the relationship between memory and trauma. For immigrants, particularly those who are refugees, the place where their past lives took place are deeply associated with traumatic memories.

But even when that is not the case, having immigrated to another country, no matter how long ago, means you are always from somewhere else. You are always betwixt and between, neither here nor there. Many important events of your life have occurred in another place. Who you are is influenced by absence; the absence of people and places that shaped those events. Distance from your original place, exile of whichever kind, necessitates understanding yourself from a different geographical and cultural setting than the one in which many of your memories were encoded. You have to organize your memories in a certain way to make them intelligible in your new context; you have to explain details that would otherwise have been taken for granted. Sometimes I feel as if I have to explain myself constantly in order to be understood by those who matter to me in this new life and place.

The *place* in which remembered events occurred is of paramount importance to immigrants. The immigrant's preoccupations are frequently focused on the vicissitudes of place and geography. This preoccupation has two components. First, whatever events may be occurring at a distance in the country of origin give it a sense of reality because of its psychological presence for the immigrant. This, I believe, is a manifestation of what Parkes (1975) described in his classical studies of grief as the "urge to search for and find the lost [love] object". Then there is for immigrants a persistent preoccupation with "what could have been" that takes different shapes: preoccupation with what could have been if the immigrants had not left their countries is a central theme in immigrants' emotional lives and ruminations about what life might have been if the immigrant had remained in her homeland or migrated to a different country or if the immigration had taken place at another life stage. She asks herself what could have happened in her life had she stayed in her country of origin. Then, her concern shifts to ask what has been gained by the migration. Finally, distortions of memory of places where events occurred develop progressively, mostly outside of conscious awareness.

My own reflections following a return to my homeland after several decades of absence manifest these preoccupations.

> My trip to Cuba made me realize that for years I had felt as if my memories had no geography. It made me realize that what I remembered had actually happened in a definite physical space that continues to exist in reality and not only in my memory. That Cuba exists beyond what I think or feel or remember about her. This realization, which may seem all too obvious, was the more powerful because before this return trip I never knew that I felt as if my country did not have a real existence beyond my memory. (Espín, 1992, p. 16)

In the interpersonal realm,

> my trip put me in touch with childhood friends and made me reflect about the differences in our lives, about the choices to stay or leave that have dramatically influenced our life projects. None of us has any way of knowing what our lives would have been like without the historical dislocations that have marked them.... It is impossible to know if our decisions have resulted in a better life project for any of us, although we each hope and believe to have made the best decision. (Espín, 1992, p. 17)

One researcher has characterized these ruminations by describing how "the migration experience creates an emergent phenomenology of incessant reference group comparisons and trade-offs between the benefits of the host society and the losses incurred in departing from the society of origin" (Rogler, 1994, p. 704). Cuban-American writer Achy Obejas, describing her own ruminations about her migration, illustrated these concerns and their relationship to sexuality in an autobiographical story. She pondered,

> What if we'd stayed? What if we'd never left Cuba?... I wonder, if we'd stayed then who, if anyone, would have been my blond lovers, or any kind of lovers at all.... I try to imagine who I would have been... but

I can't. I can only think of variations of who I am, not who I might have been. (1994, pp. 124–125)

The nagging of "what could have been" is always in the back of my mind since, as I said earlier, I don't know another me than the one I have become. All I know is that this me I am would have never taken shape had I not boarded that plane in 1961, leaving my country behind.

Place, gender and identity

As British psychologist Stephanie Taylor (2010) states in her studies of narratives of identity and place, changes in those areas have particularly impacted women in contemporary society; therefore, the burden of creating an identity is greater for women than for men. This differential impact of gender becomes even more evident when we focus on immigrant women. Regardless of age, race or social class, gender differences are present in immigrants' experiences (Espín, 1999; 2006). "Diasporic experiences are always gendered" (Clifford, 1994, p. 313). At each step of the migration process, women and men encounter different experiences. Women's roles and sexual behavior may be modified more dramatically and profoundly than men's, as a consequence of the acculturation to a new society (Espín, 1999; 2006). Young women from "racialized" groups may confront additional conflicts concerning their sexuality and body image. They have to find a balance between the imposed hypersexualization of immigrant women as "exotic" and the "hyperpurity" expected of them by their families and communities.

A generalized assumption is that women immigrants are focused on their families and homes. And, needless to say, "home" is a powerful place. "There is a long tradition of feminist criticism of the family life conventionally associated with home" (Taylor, 2010, p. 45). However, some of those critiques seem to disappear the moment we are referring to immigrant women. It seems easy to forget that gender is a social construct, and as such, it is being constructed by immigrant women in the process of remembering the old and confronting the new. Immigrant women are experiencing processes of transformation in many aspects of their lives. And while it is inappropriate to push someone beyond what they are willing to transform in themselves, it will be equally restraining to assume that immigrant women's gender roles are frozen in the past and that they are incapable of analysis and transformation concerning the ongoing performance of their gender and identity. "[A]ny identity positioning is not once and for all because social life and identity are always in the process of being lived out, in the ongoing, ever changing moment" (Taylor, 2010, p. 49). This is true of all women, including immigrants, despite the reality of the extensive self-surveillance and discipline normatively required of them by their communities that establish limits to the range of identities that can take place.

Evidently, relevant identities are never that of "a woman" but a woman of a certain class, ethnicity, age, religion and migration history. Gender is not separable from other identities. We know that the unsettled nature of social life, particularly for immigrants undergoing a process of acculturation and working through memories of the past, requires repeated performances of all identities (Taylor, 2010).

For women immigrants, the stress of gender expectations in home and society are added to the stress created by migration. Inconsistencies are unavoidable. No matter how fragile memories about women in their country of origin might be, they are the thread that tethers the immigrant to who she is; they serve to ground her both in the past and in the new place.

Loss and migration

An additional dynamic of utmost importance in the experiences and memories of immigrants and refugees relates to issues of loss, grief and mourning. Discussions of the psychological distress experienced by immigrants and refugees usually emphasize their need to cope with new experiences rather than the loss of the old familiar environment implied in the loss of home country and loved ones.

> Losses usually stem from the traumas of the uprooting experience through involuntary or voluntary extraction from the primary group networks in the society of origin. The loss of the social circle composed of intimate face-to-face contact with family, friends and neighbors is pungently distressful. (Rogler, 1994, p. 704)

Scholar of urban studies Peter Marris (1974), in his classic exploration of the impact of experiences of loss and change, described the difficulties in mourning losses brought on by ambivalent experiences such as migration, in which positive and negative outcomes overlap. According to him, "loss disrupts our ability to find meaning in experience, and grief represents the struggle to retrieve this sense of meaning when circumstances have bewildered or betrayed it" (p. 147). "The degree of choice or necessity at the time of and preparation for relocation influence the kind of loss that is experienced...and play a major role in the kind of mourning process required" (Levy-Warren, 1987, p. 307). Regardless of the conditions that determine the decision to leave, "acts of mourning attenuate the leave-taking" (Marris, 1974, p. 84). However, most immigrants are supposed to be happy that they have succeeded in migrating rather than sad for what they have lost. Sadness and grief are easily perceived as ingratitude by the members of the host culture and perhaps as incongruent even by the immigrant herself. Regardless of gratitude for their new country and relief from whatever anxieties were experienced in the home country, the transitions created by immigration demand that one grieve for the old attachments to country and people. In other words, the mixture of positive and negative feelings involved in migration complicates dramatically the emotional picture of the immigrant.

Frequently, well-meaning but under-informed friends and social service agencies emphasize the woman's adaptation to her new life and ignore her feelings of loss, despite the fact that these feelings and the need to mourn those losses are at the crux of the process of successful adaptation. These feelings of loss must become integrated into the individual's psyche. But, in so doing, identity is unavoidably transformed, and this transformation brings with it the additional need to mourn the old identity.

The loss of some of life's small, taken-for-granted pleasures may acquire enormous proportions in the emotional life of the immigrant. The pain of uprootedness is activated in subtle forms by the everyday absence of familiar smells, familiar foods and familiar routines for doing the small tasks of daily life. It is the lack of "the average expectable environment" that can become a constant reminder of what is not there anymore. It can be most disorienting and most disruptive of the immigrant's previously established identity (Espín, 1992; 1997).

Normally, the grieving process involves a moderate level of emotional disorganization. It may manifest as apathy, insomnia, loss of appetite, irritability, angry outbursts, psychosomatic symptoms and other signs of distress. When these feelings are inhibited because the loss is denied or otherwise defended against, the normal signs can become pathological by prolongation or exaggeration. Parkes's 1975 seminal study of grief and bereavement suggested several identifiable features of grief and mourning, which include a gradual process from denial to recognition and acceptance, alarm reactions such as anxiety and other related physiological symptoms, an urge to search and find the lost object, anger and guilt, feelings of internal loss of self and identification with the lost [love] object. Despite the fact that the migration may have been motivated by less than optimal conditions in the homeland, the unrealistic feeling that one has lost a paradise may persist. Idealization of the home country may increase the feeling of loss.

To compensate for these losses and feelings of grief, many immigrants struggle to maintain contact with the home country. They do this through food, music or physical proximity to other immigrants from the home country. The presence of others who share the same ethnicity and sense of cultural heritage has been found to be positively related to mental health and well-being among immigrants (Portes & Rumbaut, 1996).

Although there are similarities between the mourning of bereavement and the losses experienced with migration, the two processes differ in substantial ways. The bereaved's grief can be traced to the nature of the relationship to a specific person. While in migration, the lost object is vague and the loss pervasive. Migrants have lost country, culture and loved ones. With most of their habitual patterns being disrupted while they simultaneously learn new ones, the distress can be considerable. Immigrants themselves seldom understand the magnitude of this loss while they are experiencing it. They and others may miss the extent of its impact.

Sometimes immigrants experience reactivated grief when they retire from their jobs after many years of residing in the new country. Depression and emotional distress in later life could be traced directly to their lack of successful mourning years earlier at the time the migration took place. The urgency of survival and adaptation needs at the moment of migration forces postponing their feelings of loss and may lead to delayed reactions of grief.

The healing power of narrating memories

How can memory help to integrate all that we are into a fuller sense of our fragile selves? Well, of course, a fuller sense can be actualized through the memory work that constitutes the essence of the process of therapy but also through the telling and re-telling of the memory to others, as well as in our private and public writing.

There is much to be learned from the process of elaborating an individual narrative. For the last few years I have been engaged in the experiential analysis of my own migration story as a research project as well as in the writing of memoir about my childhood and adolescence in Cuba. My individual memories and the collective interpretations of Cuban history intermingle in my life. As I stated earlier, I am conscious of how historical events and individual lives intersect to create autobiographical memory. Now, more than 50 years after emigration and after three return visits to the places of my childhood and adolescence, not only do I *know* that leaving Cuba changed the course of my life, but also I cannot avoid wondering what my life would have been in Cuba, even if there had never been a revolution.

Narrating the events of my life and my memories of them is further complicated because Cuba is so conflictual and because so many people have opinions about it and about Cubans living abroad. I frequently encounter people who question my life decisions and judge who I must be on the basis of how I fit their own political opinions. Too frequently I feel like saying to those who question me from their self-constructed high moral ground, *What do you really know? My life experiences have to do with the powerlessness of being a toy of history, not with some interesting political theories or treatises you may have read.* And yet, despite the incomprehension I frequently encounter, perhaps because of the lack of continuity of people and places that Allende describes, I have a desperate need to talk, to tell, to show, to make myself understood. How do I explain to people who matter deeply to me but who have lived different lives what the essence of my life is? How do I explain that, in many ways, I am not who I seem to be? In other words, how can narrating my memories help to integrate a fuller sense of myself?

Memory, narratives and health

In 2001 the American Psychological Association's *Monitor* featured an article entitled *A new reason for keeping a diary*. The article reviews the results of several studies on the beneficial health effects of writing down our memories. These studies explored further "the well-established connection between writing and health" (Carpenter, 2001, p. 70). Apparently, "repeated writing about negative events decreases their emotional impact" (p. 70). Accordingly, the results of these studies demonstrate that "expressive writing – that is, writing that includes emotional descriptions of life events – helps people simplify and organize fragmented memories" (p. 69). "It squashes intrusive and avoidant thoughts about negative events as it may lead to improvements in working memory" (p. 70).

A volume edited by Lepore and Smyth (2002) with the suggestive title *The writing cure: How expressive writing promotes health and emotional well-being* presents the results of multiple studies on the effects of writing on illness such as blood pressure and cancer, as well as on a variety of emotional conditions. The studies reported in this book amply confirm the positive effects of writing narratives on populations of children and adults. They demonstrate the power of translating emotional experience into language and writing. The beneficial health effects of writing and other forms of disclosure are now a well-established fact (e.g. Jourard, 1958; 1959; Pennebaker, 1990; 1993; 1997; Lepore & Smyth, 2002). Consistently, researchers have found that writing and self-disclosure, particularly about emotionally laden events and negative experiences, improves our sense of well-being, boosts the immune system, increases our T-cell growth and antibody response, lowers our heart rate, helps us lose weight, improves sleep, elevates our mood and can even reduce physical pain, in addition to strengthening social relationships, which in turn has beneficial health effects (e.g. Carpenter, 2001; Lepore & Smyth, 2002; Pennebaker, 1990; 1993; 1997).

According to a number of studies reviewed by Norman Anderson (2003) in his book on psychology and aging, writing also appears to prolong life, even when it is only writing for yourself. It seems that writing creates changes in the way people interpret events and gives meaning to those events. When we reflect about causes and effects of life events, particularly when we do that in writing, our lives make more sense and become more manageable. The negative events become less intrusive, leaving us with more "memory space" and more energy to live our lives (Pennebaker, 1990; 1993; 1997).

For as long as I can remember, almost as soon as I learned to write, I have kept some form of diary/journal. Through the years, volumes have accumulated. These studies about the value of writing memories have been a delightful discovery. Apparently, that pile of what may have seemed more or less useless writing has played a role in helping me manage the chaotic events of my life and may have even helped to prolong it!

Because a memory that is communicated in a story shows at least a minimum of coherence, writing down memories helps the individual construct a sense of the past, even if the story is told only to oneself in the writing of a journal. The narrative communication required in the writing or telling of memories becomes the form in which the individual constructs a sense of the past and of personal continuity. Reflecting on our memories and writing about them increases critical insight and engagement. Telling our stories reinforces social bonding, learning and the memories themselves. According to Pennebaker (1990; 1993; 1997), this sharpening of communicative intent provides a mental frame of reference and a point of comparison for future decisions and actions that serves to modulate emotional reactivity and, in turn, has positive effects on health. In other words, sharing the intimate details of our lives has many functions. The act makes us feel connected to others, alleviates stress and makes us healthier.

Stories are essential because they permit moments of reflection without which actions and judgment would not be possible. In essence, in writing about experiences, the mind "is telling itself a story" that serves to create our sense of self and identity. As Donald Polkinghorne (1988) puts it, "the self needs a story in order to be" (p. 105).

Political impact of memory narratives

Life stories, although deeply personal, also have important political purposes that may also produce psychological consequences. They allow us to reinsert ourselves into the narrative that is history, to become a part of the public world by participating in the process of its making. As British sociologist Ken Plummer (1995) tells us in his book *Telling Sexual Stories*, "[s]tory telling flows in the stream of power.... The power to tell a story, or indeed to not tell a story... is part of the political process" (p. 26).

"In the United States, a country of immigrants, being an immigrant still often means to stand as a silent outsider in relation to both the culture of origin and US culture" (Giunta, 2002, p. 71). Narrating memories, either verbally or in writing, gives voice to the immigrant experience. "Memory frequently represents the inaccessible, the unspeakable" (p. 119), which is why "gathering the fragments of one's life into a unified, cohesive narrative" (p. 120) is so important "for those who have been marginalized and denied access to public forums because of their gender, race, ethnicity, nationality, language, religion, sexuality or class" (p. 120). Therefore, "memory has enormous political importance in the contemporary context because it creates or helps sharpen a much need political awareness of social issues" (p. 120) for both immigrants and others who interact with them. Appropriating their memories and narrating them is thus a powerful tool for women immigrants.

In fact, the process of valuing our memories and telling them in writing provides an important vehicle of expression to progressive individuals, feminists, clinicians, academics, students, interdisciplinary scholars and writers because it entails having the courage to challenge our old categories and make space for contradictions and ambiguities while opening us up to our new creative possibilities. When women of different generations, ethnicities or sexual/affectional orientations write and speak about their memories, we can examine the influences of age cohorts, historical generations, cultural communities or feminist priorities. Writing memories evokes unexpected varieties of feelings and relationships, expanding the definition of women's lives and identities.

Language and memory

An important consideration in the deliberate transmission of any message is the decision about what language to use for the narration. And if the construction of memories is essential for the development of identity, the linguistic decisions involved cannot be underestimated, particularly for immigrants whose memories may have been encoded in languages other than English. "You can never sidestep the question of identity when you learn to live in a new language" as Isabelle de Courtivron (2003, p. 4) says in the new collection of writings by bilingual authors she recently published. In her words, "where does the deepest material of the self, lodge itself if not in language?... [T]he elusive search for oneness, and the haunting quest for the self are perhaps foregrounded more acutely in texts by bilinguals because their authors face an ultimate disconnection. How much more difficult the fragmentation when you don't quite have 'the words to say it'?" (p. 4).

> Despite the fashionable postmodern emphasis on displacement and dislocation; despite the intellectual persuasion that trying to find wholeness in our lives is a somewhat obsolete ideal, the anxiety about fragmentation and the search for existential coherence remain primordial human responses. The life-long struggle to reconcile the different pieces of the identity puzzle (or at least to acknowledge that they cannot be reconciled) continues to be a painful and constantly renegotiated process. All the more so when the fragmentation exists in that most intimate of sites–language. (de Courtivron, 2003, p. 2)

For some people, such as me, a language other than English may be the depository of most of those memories, at once trivial and significant, of songs, poetry, places, customs and daily events, shared through many years with others who have since disappeared from my life or stayed behind in places where I am not. I experience a constant need to retell because those who are with me in this here and now don't know, don't understand, don't see. How could they? Their memory is of TV programs, sports events and popular songs that are not mine. That is why I always lose at "popular culture" and "sports" when I try to guess these categories while watching Jeopardy on TV during dinner.

In a study of autobiographical memory among bilinguals, Javier and his collaborators (1993) posited that "the nature of bilingual memory is influenced by the kind of linguistic organization the individual develops [because] memories of personal events are linguistically organized in bilinguals" (Javier, Barroso & Muñoz, 1993, p. 336). In other words, the way memories of personal events are organized in the brains of bilinguals differs from monolingual individuals – a point that Albert and Obler (1978) have also made. According to Javier, Barroso and Muñoz (1993), "the communication of memories of personal events is qualitatively different in the two languages" (p. 334); the experience is remembered differently in each language.

Examination of work in experimental psychology on bilingual autobiographical memory and clinical case reports from psychoanalytic therapy with bilinguals suggests that memory associations and retrievals for events of childhood and youth (in the country of origin) are more numerous, more detailed and more emotionally marked when remembering is done in the first language ("mother tongue") rather than in the second language (Altman, Schrauf & Walters, 2013; Schrauf, 2000).

In addition, "language serves a strong function as a retrieval cue when eliciting memories for past life experiences.... [L]anguage is specifically tied to memory traces and those traces appear to carry 'language tags'" (Altarriba, 2003, p. 316), strangely enough, even for people who do not speak that language any longer.

Moreover, it appears that "how and when the two languages are learned will determine the nature of the linguistic organization possible" (Javier, Barroso & Muñoz, 1993, p. 322). As Hoffman (2003) astutely puts it, "the kind of relationship one develops with an acquired language is deeply influenced by the kind of bond one has with one's mother or father tongue – and by extension, with all the intimacies and intimate sensations of early life" (p. 52).

Memories appear to be loaded in favor of the language in which the experience took place. In other words, "the language of the experience may not necessarily be (the bilingual person's) primary language... [but rather] the linguistic context in which the verbal interaction occurs" (Javier, Barroso & Muñoz, 1993, p. 335).

These studies' results have obvious implications for immigrant women's experiences in therapy. At best, it appears that decoding the affective meanings of memories through the use of another language is problematic. The use of English in therapy, rather than the mother tongue, may render unavailable to the therapeutic process certain areas of the intrapsychic world because apparently "memory storage is pattern storage, and patterns evoke other patterns.... [A]wareness of a small portion can trigger awareness of a whole; a single moment, a sound, or a sensation can evoke recall of an entire event" (Aragno, 1996, p. 32). Indeed, "the retrieval of information from memory is both activated and constrained by language" (Schrauf, Pavlenko & Dewale, 2003, p. 228) because both positive and negative experiences are powerfully connected to language (Dewale, 2013a, b).

> Psychotherapists working with bilingual clients often find that experience is available in some privileged way in one language versus the other. The immigrant's memories of childhood or adolescence spent in their home country may be more numerous, detailed or emotionally charged when dealt with in the mother tongue than the second language.... Representations of self-identity and the narration of emotional experience may be particularly dependent on linguistic and cultural factors.... Memories in two languages may undergird the complex sense of identity so often reported by bilinguals. (Schrauf, 2003, p. 239)

Schrauf goes on to report in several of his studies that "language is involved in both encoding and retrieval of bilinguals' memories" (p. 240). And, indeed, his results confirmed that "memories in Spanish commemorated events from before immigration; memories in English commemorated events after immigration" (p. 240).

In another study Schrauf, Pavlenko and Dewale (2003) theorize that "the argument that language might play a critical role in conditioning and constraining the path of... retrieval rests also on the notion that language is a privileged carrier of cultural meaning" (p. 239).

While it is true that psychodynamically oriented therapies pay special attention to memory processes and language associations (e.g. Amati-Mehler, Argentieri & Canestri, 1993), memory and language are undeniably essential for all forms of therapy regardless of the specific theoretical orientation. All forms of therapy are forms of a "talking cure" because language is absolutely necessary to establish rapport and to communicate during treatment. Immigrant women in therapy are involved in a process that evokes their memories and their languages in ways that question or reaffirm their sense of identity and connection with their communities. The ways in which they understand and narrate their experiences are profoundly tied to the cultures and languages of both the old and new communities.

Providing culturally relevant mental health services to immigrant women in the US unavoidably involves language use. Both multilingual clients and multilingual therapists report the significance of language in effective treatment (e.g. Dewale, 2013 a & b).

Language and identity

Language – the forced learning of the new and the loss of the old linguistic community – is central to the migration experience; it becomes evident that language changes are critical in the transformation of identity.

In fact, language change is one of the most difficult problems the immigrant faces – and I am not referring to issues of vocabulary, grammar or pronunciation (Espín, 2013). Because "language determines one's knowledge of the world, of others, and of oneself, [i]t provides a basis of support for one's identity" (Grinberg & Grinberg, 1984, p. 109). Therefore, language loss and its concomitant sense of identity loss and transformation are one of the most powerful components of the immigrant experience.

In her autobiographical account of migration, *Lost in Translation*, writer Eva Hoffman (1989) vividly describes the intensity of this experience for immigrants:

> Linguistic dispossession is…close to the dispossession of one's self.…[There is feeling that] this language is beginning to invent another me.…[And] there is, of course, the constraint and the self-consciousness of an accent that I hear but cannot control. (p. 121)

Beyond allowing the immigrants to function in the new context, a new language has profound impact on their sense of self and identity as Hoffman's statement illustrates. Like Hoffman (1989) and Giunta (2002), I am perennially homeless in language. Besides the physical distance from my place of birth, my exile is also linguistic. I will never have the same facility with language that I enjoyed the first decades of my life. Never again will I be fully comfortable in any of the languages I speak because I am not fully me in only one of those languages anymore.

Immigrants learn "to live in two languages;" similarly, we learn to live in two social worlds. Learning to "live" in a new language is not merely an instrumental process. It is not a neutral act. It implies becoming immersed in the power relations of the specific culture that speaks the specific language. Paradoxically, learning the language of the host society implies learning one's place in the structures of social inequality.

For adults, to speak with a foreign accent places one in a subordinate position within those power relations.

For children, for whom immigration usually implies schooling in a language other than the language of their parents, this process involves a "creation" of their incipient identities as members of a second-class group in the new country (Espín, 1999; 2006; 2013).

The issue becomes further complicated when different generations within a family have different levels of proficiency in the different languages spoken. While the first language or mother tongue may be taken to mean the native language of the family, in immigrant families, children are usually more fluent in the language of the host culture, which is really their first language rather than the language their parents speak (Bammer, 1994). Parents and children often end up with different native languages (Bammer, 1994). When parents and children are fluent in different languages, they may in fact be guided by different cultural codes.

On the positive side, learning a new language provides the immigrant with the opportunity to "create a new self". This facilitates working through early intrapsychic conflicts and finding new ways of self-expression that may not have been available in the world of the first language (Espín, 1999; 2013).

People who learn to use two languages have two symbols for every object. Thus, from an early age they become emancipated from linguistic symbols – from the concreteness, arbitrariness, and "tyranny" of words – developing analytic abilities to think in terms more independent of the actual word. By contrast, monolinguals may be at a disadvantage (Portes & Rumbaut, 1996, pp. 200–201).

Clearly, access to more than one language pushes at the boundaries of what is "sayable" or "tellable". While the first language is usually the "language of emotion" even for individuals who are fluently bilingual, in some cases the second language could become a vehicle for expressing concerns that could be too embarrassing to verbalize in the first language. Since the use of language is essential in counseling and psychotherapy, understanding its many implications for the immigrant woman client is of utmost importance (Espín, 1999; 2006; 2013).

A therapist who speaks the same languages of the woman immigrant can provide a unique link to her psyche – but also can a therapist who is sensitive to and aware of the importance of language even though s/he may not speak the native language of the client. In these instances, other alternatives could be used, such as having the client tell the story of an important (perhaps traumatic) life event in their native language, even if the therapist doesn't understand, as a way to access the emotional valence of the story telling, facial expressions, and so on. Pérez Foster (1998) suggests a process that I have also used since the early 1970s. Basically, the client speaks in her first language while the therapist observes emotional expression and body language. This is followed by the client providing a brief summary of what was said.

An important implication of this discussion is the need to consider language issues in therapy with immigrant women even though they may be fluent in their second language.

Obviously, some of these issues are relevant for therapy with multilingual individuals who may not be immigrants. But it is important to remember that language plays a unique part in the daily lives of women immigrants and that they are likely to have memories and associations tangled with language issues perhaps in deeper ways than those whose histories do not include the traumas of immigration.

Memory, language and ethics

Other issues that deserve attention refer to therapists' reactions to multilingual clients, the ethics of work with these populations and the need to educate therapists and their supervisors on the importance of language for these clients. Even when therapists speak more than one language, their training may not provide support for working in languages other than English. Schwartz and her collaborators (2010) address the ethical conundrums of supervising students who are providing therapeutic services in languages the supervisor does not understand.

Pérez Foster (1998) speaks of the effect on the therapy process of therapists' countertransferential reactions to persons from other cultures or to clients who speak other languages. Therapists and their supervisors may have negative associations with languages they do not understand.

Beyond language issues, memories present us in life and in clinical work with important ethical considerations. When does remembering events of individual life or collective history lead to positive outcomes or significant transformative results, and when do these memories paralyze or even destroy lives and societies? When does the duty or the need to remember become a weapon of destruction (e.g. Margalit, 2002)?

Published memoirs and narratives of trauma and conflict can be a tool of accusation and revenge as much as apology or justification. It is difficult at times to avoid making them into expressions of resentment or guilt, to develop a testimony that is not just the self-affirming monologue of a victim or a condemnation of a real or imagined executioner. They will be instruments of healing in the measure in which they can avoid these pitfalls, although in private writing, it can be necessary to process the negative feelings before psychological clarity is achieved.

If groups in situations of intractable conflict are to reach some degree of reconciliation, they must work through their unresolved pain and anger related to the memory of past intergroup encounters (Bar-On & Kassem, 2004). "Working-through enables people who have suffered traumatic social experiences to learn to live with those painful events while developing an ability to listen to the pain of the 'other.' The storytelling approach focuses on the way personal storytelling facilitates the working-through process in intractable conflicts" (Bar-On & Kassem, 2004, p. 289). The emphasis is on the working-through: on memory and narrative being used as tools for healing rather than as weapons against the other. Although Bar-On and Kassem (2004) focus on situations of group conflicts, their points are equally applicable to situations of individual trauma and any other form of individual suffering.

Some concluding thoughts

Psychologists are interested in the development of lives. Our access to lives is through stories about them. We know that life stories heard by psychotherapists are influenced by what is culturally and historically acceptable in a society because collective memories inform individual autobiographical memories. And collective and autobiographical memories can become sites of contention and conflict as well as sources of identity and self-understanding. Although this is true for everyone, the specific life experiences of immigrants make these processes more poignant.

Understanding the importance of memories of place and life events is central to the psychology of immigrant women. In addition, there are lessons here for therapists working with clients who have been dislocated and displaced in other ways. These are people who have not crossed national borders, but who are "immigrants" of a different kind, living in new regions or navigating other transitions and losses related to race, class, ethnicity, age, culture, class and sexuality as well as second and third generation women who may have been impacted by intergenerational transmission of the migration trauma. Language, place and space are also important to explore in the therapeutic context for these women who are not immigrants in the usual sense.

For psychologists whose primary interest is research, issues of memory, language, place and gender in the lives of immigrant women present an exciting wealth of ideas and opportunities for research and theorizing to investigate these processes. Once again, these studies and such theorizing may also impact lives of others who may not be immigrants to a new country but may have been displaced inside their own country.

References

Albert, M.L. & Obler, L.K. (1978). *The bilingual brain: Neuropsychological and neurolinguistic aspects of bilingualism.* New York: Academic Press.

Allende, I. (2003). *My invented country: A nostalgic journey through Chile.* San Francisco: Harper Collins.

Altman, C., Schrauf, R.W. & Walters, J. (2013). Crossovers and codeswitching in investigation of immigrant autobiographical memory. In J. Altarriba & L. Isurin (Eds.), *Memory, language, and bilingualism: Theoretical and applied approaches* (pp. 211–235). Cambridge, UK: Cambridge University Press.

Altarriba, J. (2003). Does *cariño* equal "liking"? A theoretical approach to conceptual nonequivalence between languages. *The International Journal of Bilingualism, 7*(3), pp. 305–322.

Amati-Mehler, J., Argentieri, S. & Canestri, J. (1993). *The Babel of the unconscious: Mother tongue and foreign languages in the psychoanalytic dimension.* (trans. J. Whitelaw-Cucco). Madison, CT: International University Press.

Aragno, A. & Schlachet, P.J. (1996). Accessibility of early experience through the language of origin: A theoretical integration. *Psychoanalytic Psychology, 13*(1), pp. 23–34.

Anderson, N.B. (2003). *Emotional longevity.* New York: Penguin Books.

Bammer, A. (1994). Mother tongues and other strangers: Writing "family" across cultural divides. In A. Bammer (Ed.), *Displacements: Cultural identities in question* (pp. 90–109). Bloomington, IN: Indiana University Press.

Bar-On, D. & Kassem, F. (2004). Storytelling as a way to work through intractable conflicts: The German-Jewish experience and its relevance to the Palestinian-Israeli conflict. *Journal of Social Issues, 60*(2), pp. 289–306.

Boszormenyi–Nagy, I. & Spark, C.M. (1973). *Invisible loyalties: Reciprocity in intergenerational family therapy.* New York: Harper & Row.

Carpenter, S. (2001, September). A new reason for keeping a diary. *Monitor on Psychology,* pp. 68–70.

Clifford, J. (1994). Diasporas. *Cultural Anthropology, 9*(3), pp. 302–338.

de Courtivron, I. (Ed.). (2003). *Lives in translation: Bilingual writers on identity and creativity.* New York: Palgrave Macmillan.

Dewale, J.-M. (2013a). *Emotion in multiple languages.* (2nd ed). Basingstoke, UK: Palgrave Macmillan.

Dewale, J.-M. (2013b). Multilingual clients' experience of psychotherapy. *Language and Psychoanalysis, 2*(2), pp. 31–50.

Espín, O.M. (1992). Roots uprooted: The psychological impact of historical/political dislocation. In E. Cole, O.M. Espín & E. Rothblum (Eds.) *Refugee women and their mental health: Shattered societies, shattered lives* (pp. 9–20). New York: Haworth Press.

Espín, O.M. (1997). *Latina realities: essays on healing, migration, and sexuality* (pp. 19–30). Boulder, CO: Westview Press.

Espín, O.M. (1999). *Women crossing boundaries: A psychology of immigration and the transformations of sexuality.* New York: Routledge.

Espín, O.M. (2006). Gender, sexuality, language, and migration. In R. Mahalingam (Ed.), *Cultural psychology of immigrants* (pp. 241–258). Mahwah, NJ: Lawrence Erlbaum.

Espín, O.M. (2013). "Making Love in English" language in psychotherapy with immigrant women. *Women and Therapy, 36*(3/4), pp. 198–218.

Foote, K. (1993). *Shadowed ground: America's landscapes of violence and tragedy*. Austin, TX: University of Texas Press.

Giunta, E. (2002). *Writing with an accent*. New York: Palgrave Macmillan.

Grinberg, L. & Grinberg, R. (1984). *Psicoanálisis de la migración y del exilio*. Madrid: Alianza Editorial [English translation by N. Festinger. *Psychoanalytic perspectives on migration and exile*. New Haven, CT: Yale University Press, 1989].

Halbwachs, M. (1980). *The collective memory*. New York: Harper & Row Colophon Books. [*La mémoire collective*. Paris: PUF, 1950].

Halbwachs, M. (1992). *On collective memory*. (Trans. & Ed. Lewis A. Coser). Chicago: University of Chicago Press. [From: *Les cadres sociaux de la mémoire*. Paris: PUF, 1953 and *La topographie légendaire des évangiles en Terre Sainte: Etude de mémoire collective*. Paris: PUF, 1941].

Hartman, S., Hoffman, E., Mendelsohn, D. & Miller, N.K. (2011). Memoirs of return. In M. Hirsch & N.K. Miller (Eds.), *Rites of return: Diaspora poetics and the politics of memory* (pp. 107–123). New York: Columbia University Press.

Hoffman, E. (1989). *Lost in translation*. New York: Dutton.

Hoffman, E. (2003). P.S. In I. de Courtivron (Ed.), *Lives in translation: Bilingual writers on identity and creativity* (pp. 49–54). New York: Palgrave Macmillan.

Hobsbawm, E. & Ranger, T. (1992). *Invention of tradition*. New York: Cambridge University Press.

Ignatieff, M. (1996). Articles of faith. In U. Owen (Ed.) *Wounded nations, broken lives: Truth commissions and war tribunals*. Retrieved February 26, 2004, from the Index of Censorship, 5/96 website: http://www.oneworld.org/index_oc/issues5/96/ignatieff.html

Javier, R., Barroso, F. & Muñoz, M. (1993). Autobiographical memory in bilinguals. *Journal of Psycholinguistic Research, 22*(3), pp. 319–338.

Jelin, Elizabeth (2003). *State repression and the labors of memory*. (Trans. J. Rein & M. Godoy-Anativia). Minneapolis, MN: University of Minnesota Press. [*Los trabajos de la memoria*. Siglo XXI, 2002].

Jourard, S.M. (1958). A study of self-disclosure. *Scientific American, 198*(5), pp. 77–82.

Jourard, S.M. (1959). Healthy personality and self-disclosure. *Mental Hygiene, 43*(4), pp. 499–507.

Kandel, E.R. (2006). *In search of memory: The emergence of a new science of mind*. New York: Norton.

Lepore, S.J. & Smyth, J.M. (2002). *The writing cure: How expressive writing promotes health and emotional well-being*. Washington, DC: American Psychological Association.

Levy–Warren, M.H. (1987). Moving to a new culture: Cultural identity, loss, and mourning. In J. Bloom–Feshbach & S. Bloom–Feshbach (Eds.), *The psychology of separation and loss: Perspectives on development, life transitions, and clinical practice* (pp. 300–315). San Francisco: Jossey–Bass.

Margalit, A. (2002). *The ethics of memory*. Cambridge, MA: Harvard University Press.

Marris, P. (1974). *Loss and change*. London: Routledge and Kegan Paul.

Obejas, A. (1994). *We came all the way from Cuba so you could dress like this?* Pittsburgh, PA: Cleis Press.

Ochberg, R.L. (1992). Social insight and psychological liberation. In G.C. Rosenwald & R.L. Ochberg (Eds.), *Storied lives: The cultural politics of self-understanding* (pp. 214–230). New Haven, CT: Yale University Press.

Pennebaker, J.W. (1990). *Opening up: The healing power of confiding in others.* New York: Morrow.

Pennebaker, J.W. (1993). Putting stress into words: Health, linguistics, and therapeutic implications. *Behavior Research and Therapy, 31,* pp. 539–548.

Pennebaker, J.W. (1997). Writing about emotional experiences as a therapeutic process. *Psychological Science, 8,* pp. 162–165.

Pérez Foster, R. (1998). *The power of language in the clinical process.* Northvale, NJ: Jason Aronson.

Pillemer, D. (2000). *Momentous events, vivid memories: How unforgettable moments help us understand the meaning of our lives.* Cambridge, MA: Harvard University Press.

Plummer, K. (1995). *Telling sexual stories: Power, change, and social worlds.* New York: Routledge.

Polkinghorne, D.E. (1988). *Narrative knowing and the human sciences.* Albany, NY: State University of New York Press.

Portes, A. & Rumbaut, R.G. (1996). *Immigrant America: A portrait.* (2nd ed). Berkeley, CA: University of California Press.

Rogler, L.H. (1994). International migrations: A framework for directing research. *American Psychologist, 49*(8), 701–708.

Rosenwald, G.C. (1992). Conclusion: Reflections on narrative self-understanding. In G.C. Rosenwald & R.L. Ochberg (Eds.), *Storied lives: The cultural politics of self-understanding* (pp. 265–289). New Haven, CT: Yale University Press.

Rosenwald, G. C. & Ochberg, R. L. (Eds.). (1992). *Storied lives: The cultural politics of self-understanding.* New Haven, CT: Yale University Press.

Rubin, D. (Ed.) (1986). *Autobiographical memory.* New York: Cambridge University Press.

Salomon, G. (2004). A narrative-based view of coexistence education. *Journal of Social Issues, 60*(2), pp. 273–287.

Schacter, D. L. (1996). *Searching for memory: The brain, the mind, and the past.* New York: Basic Books.

Schacter, D.L. (2001). *The seven sins of memory: How the mind forgets and remembers.* Boston: Houghton Mifflin.

Schrauf, R.W. (2000). Bilingual autobiographical memory: Experimental studies and clinical cases. *Culture and Psychology, 6,* pp. 387–417.

Schrauf, R.W. (2003). A protocol analysis of retrieval in bilingual autobiographical memory. *The International Journal of Bilingualism, 7*(3), pp. 235–256.

Schrauf, R.W., Pavlenko, A. & Dewaele, J-M. (2003). Bilingual episodic memory. *The International Journal of Bilingualism, 7*(3), pp. 221–233.

Schwartz, A., Domenech-Rodríguez, M.M., Santiago-Rivera, A., Arredondo, P. & Field, L. (2010). Cultural and linguistic competence: Welcome challenges from successful diversification. *Professional Psychology: Research and Practice, 41*(3), pp. 210–220.

Taylor, S. (2010). *Narratives of identity and place.* New York: Routledge.

2

Exploring Intersections of Privilege and Oppression for Black and Asian Immigrant and US Born Women: Reaching across the Imposed Divide

Karen L. Suyemoto and Roxanne A. Donovan

Psychologists and other social scientists are increasingly attending to intersectionality, recognizing that sociopolitical statuses interact to create qualitatively different experiences (Cole, 2009). For example, Black and Asian[1] women in the United States experience racism and sexism generally, racialized sexism, and gendered racism (Collins, 2000; Ghavami & Peplau, 2013; Moradi & DeBlaere, 2010), regardless of whether they are immigrants or US born. However, immigrant and US born women may differ in how they understand and react to these experiences, which affect the ways in which they relate to each other in response to these experiences. These differences are influenced by developmental and current experiences with power and privilege associated with nativity (and with other sociostructural variables).

In this chapter, we apply a structural power analysis to explore interpersonal divides between immigrant and US born women within Black and Asian racialized groups. We argue that tensions between immigrant and US born women in these groups are influenced by imposed meanings of gendered racialization created to maintain White male supremacy. Including both Black and Asian women enables us to examine how conflictual interactions are not necessarily inherent in specific cultural or racial dynamics of a particular minority group. We begin by presenting a brief overview of Black and Asian populations in the US and our personal positionalities, with specific attention to nativity. We identify nativity-related influences on perspectives of racialization and ethnicity and then critically examine the ways in which these perspectives and associated relational tensions relate to meanings imposed by the dominant US discourse of gendered race for Blacks and Asians. We conclude by reflecting on the challenging process of developing awareness of relative power and privilege as an immigrant or US born Black or Asian woman in order to foster greater unity. Throughout the chapter, we use vignettes from our personal and observed experiences to illustrate the divides and processes we explore.

Given the dearth of literature in this area, we view this chapter as a beginning, inevitably incomplete and tentative. There is little empirical literature examining intersections of race, gender, and nativity for Black or Asian women in the US and even less related to the ways that these intersections affect in-group relations. There has been some (albeit minimal) empirical exploration of generational tensions within families or between mothers and daughters due to differences in gendered enculturation, racialization, and internalization of dominant White European American meanings (e.g. Suh, 2007; Waters, 1999), and we do draw from this literature. However, intergenerational family dynamics are different from interpersonal dynamics between age-group peers or community members. Our analysis is therefore largely built from the integration of observations and analysis of related but not focused research. Thus, we cannot aim to describe the modal experiences of these groups or subgroups of women or the generalizable relations between variables that affect each group or interactions between groups. Our goal, instead, is to raise questions and to consider how reflection on these questions may enable us to build bridges to create empowering unity.

Our approach to this chapter rests upon certain feminist principles and epistemological stances. The emphasis on a structural power analysis is a central tenet of feminism, as is the focus on intersectional experiences. To understand the experiences of immigrant women, psychologists must examine both the shared experiences (e.g. experiences shared by immigrant women across other statuses) and those that are uniquely shaped by different intersections (such as racialization) while acknowledging that it is impossible to do full justice to all the intersectional complexities of any group of women.

Our approach is also shaped by the foundational feminist principle that the personal is political; there is a continuous interaction between individuals, groups, and the sociohistorical or sociostructural context of power and privilege (Espin & Gawelek, 1992). We utilize our personal experiences and observations as the springboard for exploration, offering these not only as illustrations of reflexivity, but also as valid data, reflecting a critical ideological philosophy and a constructivist-interpretivist epistemological stance. We do, however, also recognize the importance of reflexivity, particularly in acknowledging the limitations of and bias in any analysis, regardless of whether the basis is empirical, anecdotal, personal, or based on previous literature (Finlay, 2002). We recognize, for example, that using our own experiences means that we inevitably examine intersections and interpersonal experiences that are most salient to our particular positionalities. This enables us to describe more fully the affective as well as cognitive impacts and questions within this context. However, in limiting our relative focus we do not mean to convey any evaluation of whose experience is more important or central for consideration.

Finally, we see this chapter as part of our own journey toward *conocimiento* (Anzaldúa, 2002), a concept of emotional, cognitive, and spiritual empowered awareness and action we believe reflects feminist consciousness for women of color. We have approached this chapter as an opportunity to engage the interpersonal risks we believe are necessary to create connections across the imposed divides. Doing so entails considerable trust. In many ways, women of color have learned to deny, swallow, or compartmentalize past experiences where trust was betrayed. In writing this piece, we needed to confront those experiences and take the risk of trusting again: trusting ourselves, each other, and you as the readers. Writing from a personal space makes us feel vulnerable and exposed. However, we have learned (and we relearn regularly) that being vulnerable and exposed is one way to create change in ourselves and others.

Group and personal contexts and positions

The context of Black Americans

There are approximately 42 million Black people in the US, making up about 14 percent of the current US population (U.S. Census Bureau, 2011). Of that amount, only 9 percent are Black immigrants. The children of Black immigrants make up an additional 7 percent (U.S. Census Bureau). The remaining 84 percent are mostly descendants of African slaves or multigenerational descendants of Black immigrants who moved to the US after emancipation. A large proportion (52 percent) of first-generation Black immigrants were born in the Caribbean, particularly the Anglophone Caribbean (also known as the British West Indies) (Thomas, 2012; Waters, 1999).

Typical of this West Indian group, I, Roxanne, immigrated with my family in the late 1970s, relocating from Georgetown, Guyana to Queens, New York. I was eight years old, spoke English well and had previously visited the US, but the change was abrupt nevertheless. I went from being typical to being atypical overnight. Aside from our family's financial struggles, my biggest challenge was racialization. Similar to the rest of the Caribbean, group boundaries in Guyana are defined more by ethnicity and class than by race (Benson, 2006; Waters, 1999). Additionally, like me, many Guyanese are multiracial, with some combination of African, European, Indian/South Asian, Hispanic, and indigenous heritage. As a result, I did not think of myself in racial terms in Guyana. But in the US, I learned quickly that race mattered – to White European Americans *and* to African Americans. This message was repeatedly and clearly delivered through multiple experiences and interactions, including the numerous White European Americans who have asked "What are you?" or have tolerated my presence in white spaces because I am "not really black" and the numerous African Americans, particularly women, who have asked "Are you black?" or have been suspicious of my presence in black spaces because I am "not black enough". I remember thinking if I were just a little darker and had curlier hair, like my sister, all my problems would be solved. My internal wish, however, conflicted with the external messages I received from the dominant White European American society that being a West Indian immigrant female was somehow better than being an African American female. It was in this bifurcated state of wanting to be accepted by African Americans but not wanting to be rejected by White European Americans that I navigated my early years in the US, oblivious to the central role ascribed privilege was playing in my experiences.

The context of Asian Americans

There are approximately 17 million people of Asian descent in the US, making up 6 percent of the current US population (Hoeffel, Rastogi, Kim & Shahid, 2012). Chinese, Filipino, and South Asians are the largest specific ethnic groups; they comprise over 50 percent of the Asian population. The overwhelming majority of people of Asian descent in the US are monoracial, claiming only one race. The majority are also foreign born, US citizens, and bilingual, and almost half live in the western part of the US (Hoeffel et al., 2012; Suro et al., 2007).

As a multiracial Japanese American born and raised in the Boston metropolitan area, I, Karen, have always been aware that my experiences do not reflect those of the "typical" Asian American. I grew up with my sister in an affluent, primarily white suburb of Boston. My father was a Nisei who had been imprisoned in the camps during WWII and was therefore distant from and ashamed of his Japanese heritage. My mother was White European American from a rural southern background, and my stepmother was of Russian Jewish heritage from New England. As a result, what it meant to me to be Asian or Japanese American was initially much less cultural than it was racial, defined more by White European American people than by an Asian American or Japanese American group. I learned early the dominant White European American gendered and racialized meanings marking me as an Asian American woman. From experiences of being chased home by White male peers on the anniversary of Pearl Harbor Day in elementary school, I knew I represented something unwelcome. From being stopped on the street by a White European American man and sexually propositioned because "Asian American women are so exotic," I learned that my race was gendered and my gender racialized. Simultaneously, I learned that "Asian," at least for me, was complex and not easily understood: From experiences of being asked "What are you?" and having my answer contradicted by White *and* Asian people, I learned I did not fit into what was expected. These experiences shaped my understandings of the meaning of being Asian American and of my place within that meaning, in relation to gendered racialization. What I did not and could not see as clearly while growing up were the many relative and ascribed privileges I had in relation to other women of Asian descent in the US, and particularly in relation to those from immigrant backgrounds. It is the nature of privilege that a major unearned advantage is to not recognize privilege when you have it.

For both of us, being multiracial shaped our experiences of race, ethnicity, immigration/nativity, and relations with other women within our racialized groups. While recognizing this, we choose here to consider the multiracial experience as related to our particular positionalities within our groups rather than to make this part of our identities our primary focus. It is not uncommon for West Indian immigrant women to be multiracial; Japanese Americans have a higher proportion of multiracial individuals compared to other Asian ethnic groups; and interracial marriage is related to generations in the US. The multiracial experience is confounded with ethnicity, US nativity, and family history. Thus, we integrate the intersection of our multiracial experiences in our analysis rather than center on multiraciality.

Navigating boundaries: Perspectives on gendered racialization across nativity

During my, Roxanne's, first or second year in the US, my teacher told the class to raise our hands when our racial group was called. I had no idea what to do. In Guyana, I had never been asked to identify with a racialized group – I was Guyanese and that was all that mattered. Now I was being asked to pick a group, and Guyanese was not among the options. So, I raised my hand when the teacher said white…and when she said black…and when she said Hispanic. This did not go over well. The teacher informed me I had to choose one group. What did that even mean? To make matters worse, the other students, all US born, saw my confusion and began laughing. Even my best friend, the only other Black female in the class, laughed. I fought back tears. Although I felt alienated by my teacher and the other students, I felt especially betrayed by my best friend. I did not expect her to join in what felt like ridicule. As the only two Black girls in the class, she and I stood apart from the other mostly White students. In addition, we both lived in small apartments and were part of families that struggled financially, in contrast to most of our classmates who lived in houses and seemed well-off financially. I thought these shared experiences bonded us together, eliminating or at least minimizing differences related to our birthplaces. Her laughter ruptured this assumption and underscored that she had things in common with our classmates that I didn't even understand, further marking me as an outsider.

I, Karen, had just met my Asian American faculty colleagues. Although I had not started my new position yet, I'd been invited to participate in a curriculum and program development retreat. Four of the five were immigrant women, the fifth was a US born man who was largely silent. I was looking forward to talking with them about the new Asian American psychology course I had been hired to develop. I needed their perspectives because it was a new context for me. My new university had a large proportion of students of color, students of Asian heritage, and students with immigrant/refugee backgrounds. Within the first few minutes of the discussion, one of the women asked, "What would you say if a student said that her client was being forcibly hospitalized and she (the student) didn't know what to do?" I started out my reply by stating that the student, as a professional counselor or human service worker, would need to know that there were laws against forcible hospitalization and that the student would need to advocate for the client's release if the client were not a danger to self or other. Before I could even finish my initial reply, one of my new colleagues interrupted me, and there ensued what felt, to me, like an extensive attack where she and others informed me that I did not understand; that this student was an immigrant or refugee and would not have the cultural knowledge or privilege to advocate against the system or speak out against doctors who had more authority; that I was not considering that the student might fear deportation or other consequences from authorities if she made waves; that I would be demonstrating extreme insensitivity if I taught this; that I would be damaging students; that I took for granted the extensive privilege I had as a US born Asian American. The last few minutes of this "discussion," I was crying in front of people I had barely met. I left the room. I was in shock: what had happened? Was I so terrible, so insensitive, so stupid and unworthy? I had thought these colleagues would be the support I had not experienced in primarily White European American academic settings. I felt battered, as if what was being conveyed to me was that I would never be able to be successful in this class, with these students, teaching about or to Asian Americans or as an Asian American myself. These women, from whom I had felt so excited about learning, had clearly put me in my place – outside.

These stories illustrate the ways in which both immigrant and US born women (and girls) enact assumptions about the "other" that cause pain and confusion, contributing to creating distance. Although these assumptions are not experienced by only women, they create divisions that affect the ability of women of color to focus on shared experiences within gender and race and to unite to resist oppression. Pain, shock, or confusion are frequently initial reactions to being "othered." But often, rather than interrogate that misunderstanding, women of color retreat, internally "re-enact[ing] our past hurts" (Anzaldúa, 2002, p. 572), holding onto the ways in which the attack confirms the worst fears about ourselves and others.

Our feminist and multicultural training has taught us that interactions such as these are frequently fraught with power dynamics. What is happening here? What is being assumed, correctly or erroneously, by the immigrant or US born women? What aspects of the other's privilege are being highlighted, and what aspects of one's own privilege are being ignored?

One base disconnection between immigrant and US born women relates to differential emphases on ethnicity and ethnic cultural affiliation versus race and racialized affiliation (Ancheta, 2006; Suyemoto, 2002; Waters, 1999). West Indian and East Asian immigrants come from home countries where the dominant group is racially Black or Asian, respectively. In their home countries, they are in the racially privileged position, recipients of the many invisible daily unearned benefits and lack of microaggressions that accompany the dominant racialized status. The meaning of being a woman and the benevolent and hostile sexism experienced in home countries are not inherently racialized in intersecting oppression as they are in the US (Ghavami & Peplau, 2012). Thus, immigrant women may not initially recognize that they are affected by *racial* oppression, may not emphasize this effect, or may hold strong beliefs that oppressed racialized status can be overcome with hard work (Model, 2008; Waters, 1999). Simultaneously, immigrant Black and Asian women relatively new to the US experience culture shock and acculturative stress. Every aspect of their daily living has changed: individual, microsystem, mesosystem, and exosystem (Hong & Ham, 2001; Schwartz, Unger, Zamboanga & Szapocznik, 2010). Furthermore, they have chosen or been forced to make a life change that introduces high uncertainty as their "alien" status inherently reflects external constraints on choice and control (power). These constraints may affect whether, when, and where they are allowed to work or travel; whether they may remain in the US if their school or work or marital status changes; whether they may become a citizen if they desire; and so forth.

In contrast, US born children of all races are socialized into a racialized and racially gendered system from birth. People in the US, including US born minority women, internalize this socialization regardless of their desire or intention, as demonstrated by research on stereotype threat, implicit associations to racialized groups, aversive racism, and internalized racialism and racism (e.g. Cokley, 2002; Devos & Banaji, 2005; Dovidio, Gaertner, Kawakami & Hodson, 2002; Gupta, Szymanski & Leong, 2011; Pyke & Dang, 2003; Pyke & Johnson, 2003; Steele, Spencer & Aronson, 2012). African American and Asian American women further experience unique racialized sexism and gendered racism (Ghavami & Peplau, 2012). They are also on the margins of the margins because the prototypical image of an American woman is White and that of a Black American is male (Ghavami & Peplau, 2012; Purdie-Vaughns & Eibach, 2008; results for Asian Americans are inconsistent in relation to prototypicality and gender). Similar to immigrant women, US born Black and Asian Americans have to balance two cultures and cope with systemic European American ethnocentrism. However, the development of this balance has been a gradual process of negotiations and identity development from birth (Park, 2008) rather than an abrupt transition from and loss of previously established homeostasis (Hong & Ham, 2001).

In my example above, I, Roxanne, had no racialized lens to view my experience. My interactions during my first few years in the US were focused on trying to find some connection with the other students to reduce the isolation I was feeling. I longed to belong – to fit in like I did in Guyana. I understood the rejections I experienced as being the result of things related to being born outside of the US: things that could change, such as my accented English and not having financial resources like many around me. I did not know or understand how the US racialized system influenced people's perceptions of me and other people of color. I assumed my best friend fit in and was accepted in ways I was not because she was born in the US and went to school with the other children since kindergarten. I understood that she, like me, was marginalized because of class, but it did not occur to me that she and I were also significantly marginalized because of our racialized group membership. On a superficial level, I realized that Black people in the US were in the minority and were subject to discrimination, but I had yet to learn the depth and far-reaching nature of racism. In contrast, my best friend was well aware of classist and racist oppression; she had experienced them all of her life. These intersecting marginalizations probably hurt her deeply because, unlike in my situation, they were perpetrated by people she had known all of her life and with whom she shared a birthplace and a similar culture.

By virtue of being born in the US, my best friend was ignorant of my acculturation struggles. She likely saw me as living an adventure; it was wonderful that in Guyana I had lived in a large house by the ocean whereas she had always lived in a small apartment. I believe she also mistakenly assumed that I understood and placed myself within the racialized hierarchy she had experienced all of her life. Although she was biracial, her phenotypic presentation marked her as black, and based on the norms at the time, she was required to identify as black. As such, my identifying as White, Hispanic *and* Black may have seemed at the very least like I was trying to distance myself from our shared Black racialization, and at the very most like I was selling out her and my other Black male classmate. In a moment where I was feeling betrayed and isolated, I suspect she was feeling the same. Within these spaces of misunderstanding, hurt was able to take root and grow, creating a boundary that encased *and* separated us.

In my example, with my background as a US born Asian American, I, Karen, came to the conversation with an assumption that we were all Asian Americans, unified in our racialized experiences and resistance to racism. My uncertainty about our unity came from complex fears of being perceived as "not Asian enough" due primarily to my being multiracial (see Kinsley, 1994), but also to the lack of Japanese cultural heritage that I attributed to my father's response to the camps and therefore to racialized oppression. Although I did not know how my new colleagues felt about multiracial authenticity, I rather assumed that they would be aware of these kinds of dynamics within the racialized Asian American experience. I confess that I had given minimal thought to my nativity and citizenship privilege, to the privilege associated with my English proficiency and lack of accent; to cultural privileges I enjoyed from having familiarity with the dominant US cultural practices; or to the possibility of relative *privilege* because of my multiraciality. I did not even realize that the language "Asian American" was itself exclusive and complex, which was part of my privilege.

In contrast, I believe my colleagues came with a very different view. They assumed that I would *not* be unified with them, because I would not understand the experiences most salient to them related to the lack of cultural and citizenship privilege. I suspect they were disturbed or even offended by what they perceived as a lack of attention to these issues by US born people of East Asian heritage. Furthermore, they were either not familiar with or did not prioritize the US system of racialization in the same way I did; they seemed to have little familiarity with the complexities of being multiracial and consistently "othered" in ways that could be experienced as the antithesis of privilege, even within East Asian American communities in the US. I believe they saw me as pretty much wholly privileged in relation to them: not only was I US born, but as a Japanese American, I belonged to the only major Asian ethnic group in the US where the majority of individuals were US born. I was also East Asian, which meant that I was part of the group that was at the center of the dominant White European American construction of Asian American (Suyemoto, 2002), in contrast to South Asians or Southeast Asians. Their first thought in interacting with me was not about what it was like to be shaped by being the racialized other as a child – that is, to be defined as other by both the dominant White European American group and by Asian Americans.

In sum, US born blacks and Asians have a developmental experience of racialization that builds the understanding that there is no choice but to negotiate the imposition of denigrated racialized meanings. These experiences may relate to a relatively lesser emphasis on ethnic distinctions within racialized groups, given the emphasis on racial unity (Park, 2008). In contrast, West Indian Black and East Asian immigrant women may feel that the primary basis of their exclusion is cultural or linguistic and that the best way to move forward is through simultaneous acculturation and cultural maintenance, embracing the ideology of cultural pluralism (Deaux, Reid, Martin & Bikmen, 2006). Given that racialization and racialized categories are not a central part of many West Indian and East Asian immigrants' self-schemas, they may resist identifying racially or pan-ethnically as Black or as Asian because their specific cultural and national/nativity identities are most important (Ancheta, 2006; Waters, 1999). This may be particularly the case for immigrant women (in contrast to men), whose gender roles include conveying cultural norms and pride. Women from each nativity space are likely to approach the other with a greater focus on their own oppression rather than their own privilege and may therefore not even see the influence of their "invisible"[2] privilege (McIntosh, 1997). If they have an analysis of privilege and oppression, it is most likely to be one that focuses on the ways they feel othered by those who have privileges they lack (Vasquez & McGraw, 2005).

Although understanding the different perspectives of immigrant and US born Black and Asian women on racialization and culture may be a step toward bridging the divide, relations between immigrant and US born Black and Asian women in the US are always occurring in the context of meanings about gender, race, and nativity that are imposed by the dominant White European American male context. Deconstructing the power dynamics within these interactions further illuminates the origin and function of our assumptions about the other, and the challenges that must be overcome to foster greater unity in order to resist oppression.

The dominant White European American male context: Imposed oppressive meanings

"Are you Black?" I, Roxanne, get asked this question a lot by all kinds of people across the racialized spectrum. Growing up, this question made me feel uncomfortable and isolated, like I was somehow alien because I didn't fit neatly into a recognized and predefined black category. I also didn't know what the right answer was. What did it mean to say "yes" when I was multiracial, wasn't always perceived as black, wasn't born in the US, and received messages by some West Indian immigrants in my community that there was something bad about being Black American? What did it mean to say "no" when I had countless experiences of anti-black racism, when my national identity was no longer relevant or even understood (most White people think I'm saying Ghana when I'm saying Guyana), and when I didn't have any other appropriate label? So I hedged, usually identifying as Black-/Afro-Caribbean when interacting with White people and African Americans and usually identifying as Guyanese when interacting with people from the Caribbean. I started to question this strategy after I left New York City and, for the first time, began interacting regularly with many African American women. Some of these women shared stories of feeling rejected and pained by Black immigrants whom they perceived as overtly distancing themselves from African Americans through behaviors such as highlighting being foreign born. One of these women went a step further by saying that such behaviors made it hard for her to trust that Black immigrants would care for or support her, particularly around challenging racism. These stories were painful to hear because they raised both my sense of insecurity about being an outsider and my awareness of inadvertently perpetuating the myth of African American and black inferiority.

When I, Karen, was younger, I knew I was perceived by other East Asian heritage women as a "banana" (yellow on the outside but white on the inside). I didn't speak Japanese; I didn't know the right mannerisms; I couldn't even name (let alone eat appropriately) a lot of the food. Plus, I was multiracial, which pretty much excluded me from being "really" Asian American. I grew to understand that I would always be outside, never able to be "authentically" Asian American. (It was never an option to be simply American, given the racism I experienced from White European Americans.) Although I attributed this painful feeling of exclusion for many years to being multiracial, I came to realize that this was not a unique multiracial experience. My Asian American friends, students, and research participants from both immigrant and US born backgrounds frequently describe their own similar feelings and experiences, related to interactions between Asian immigrants and US born Asian Americans. East Asian immigrant women describe being called FOBs ("Fresh off the boat") and being avoided or ostracized by US born East Asians in classes and social gatherings, never "American" enough. US born East Asian Americans describe being called "twinkies" (the newer banana jargon), "ABCs" (American-born Chinese) or "whitewashed," criticized for being too American, never feeling authentically Asian. These stories are accompanied by confusion, hurt, and sometimes anger from both "sides," usually without a recognition that the binary framing of the identity choices means that there is no positive option for being of Asian heritage in the US.

Race is widely acknowledged as a social construct created to justify and maintain hierarchies of power and privilege (Smedley & Smedley, 2005). The dominant group has the primary influence on socially and legally defining the meaning of "Black" and "Asian." Definitions of black as slave, criminal, less intelligent, lazy, worth three-fifths of a White man, and so forth (Guthrie, 2004) are all part of the historical and current legal and social meanings of black in the US. Definitions of Asian as alien, enemy, untruthful, inferior, un-American, and so forth (Ancheta, 2006; Kim, 2006) are all part of the historical and current legal and social meanings of Asian in the US. These meanings continue to be socially disseminated, and the legacies of previous legal codification affect our institutional systems (e.g. salary disparities, health disparities, lack of representation in education, and bias in legal and court systems).

Furthermore, *gendered* racialized meanings are embedded within racism, shaping an intersectional experience for women of color. African American women are stereotyped as jezebels, matriarchs, welfare queens, and mammies (Collins, 2000; Donovan, 2011; Patton, 2001). East Asian American women are stereotyped as geishas, china dolls, dragon ladies, and tiger moms (Patton, 2001; Prasso, 2006). These stereotypes partake of *and* preserve historical constructions that maintain sexism and racism and that illuminate their intersectional nature. For example, the jezebel image of African American women portrays Black women as lustful and hypersexual and relates to historical contexts where white slave-owners justified sexual assault of Black female slaves (White, 1985). Not only is this image historical, but it also continues to influence how young Black women are viewed, maintaining their oppression through behaviors such as blaming Black female sexual assault survivors (Donovan, 2007; Donovan & Williams, 2002). There is a similar historical continuity to stereotypes of women in the Asian community. For example, the Page Law of 1875 prohibited entry into the US of Chinese, Japanese, and "Oriental" prostitutes. Any Asian immigrant women thus required proof of "good character" to enter the US. The law was rooted in a view that Asian women were essentially sexualized and dangerous to White America, akin to stereotypes of geishas and dragon ladies. Application of this law essentially excluded almost all Asian women, controlling the negative influence of Asian women's sexuality and precluding the propagation of the foreign culture (Abrams, 2005; Ancheta, 2006).

Finally, *nativity* enters this oppressive intersectional mix. In each group, the nativity majority is centralized in the denigrated focus. The modal black experience is US born, and so the dominant denigrated black experience is related to African Americans. In contrast (but with similar results to support White European American male supremacy), the modal Asian experience is immigrant, and so the dominant denigrated stereotype focuses on characteristics that are perceived as closer to heritage country values/practices and therefore as foreign. Even newer stereotypes that seem positive on the surface are subject to denigration. For example, the contemporary "positive" image of Black women as tough, hardworking and emotionally strong is framed as the result of the legacy of slavery in the US and denigrates Black women (i.e. African American women) as hypofeminine and aggressive (Romero, 2000). Similarly, the model minority stereotype is being increasingly framed in relation to the Asian "invasion" at colleges and to familial socialization and academic expectations that are foreign (e.g. the discourse about "tiger moms" with Chinese values rather than American values).

Regardless of a woman's nativity, racialized stereotypes are applied by others. However, responses of immigrant and US born women in relation to these stereotypes are different and may fuel interpersonal tensions. African American (US born) and East Asian immigrant women may feel particularly persecuted by these stereotypes because they are, indeed, explicitly targeted at them. The majority of Black women in the US, and African American women specifically, *are* descendants of US slaves; but this has no association with being hypersexual, immoral, lazy, unintelligent or subservient. Similarly, the majority of Asian women in the US, and East Asian immigrant women specifically, *are* foreign born, but this has no association with being enemy, sneaky, sexual, subservient or dangerous. As with any systemic oppression, the problem is in the stereotype, not in the women.

In contrast, West Indian immigrant and US born East Asian American women may view themselves as *exceptions* to the stereotype. West Indian immigrant women do not have ancestral ties to the legacies of slavery in the US and Asian American women are not foreign born. These women may experience feelings of confusion or believe the stereotype is being unjustifiably personally applied without recognition that a stereotype is, by its nature, unjustifiable. They may negatively judge others within the racialized group because they seem to fit the stereotype, and they may work to particularly emphasize how they are exceptions and should be treated differently (Kim, 2006; Pyke & Johnson, 2003; Rong & Fitchett, 2008; Waters, 1999).

These strategies may be felt or justified as attempts to create a positive identity in an oppressive context, but they may also be less consciously aimed at accessing ascribed or relative privilege not available to the denigrated majority. However, the cost of such strategies is internalization of oppressive racialized stereotyping (Kim, 2006; Pyke & Johnson, 2003); the nature of the exceptionality *and its limits* are created by the White European American male dominant group. Furthermore, although the strategy of exceptionality may offer some protection from discrimination through ascribed relative privilege, these benefits are accessible only if accompanied by the expected subservience. US born East Asian women may benefit from unaccented English but those who actively resist the stereotypes or who are not appropriately deferential are quickly put in their place, as are "uppity black" West Indian women.

The interpersonal tensions described in the vignettes above are influenced by the ways in which immigrant and US born women struggle with internalizing or resisting oppressive definitions imposed within the White European American male-dominant context. For African American women, resisting harmful stereotypes and developing a positive identity may entail prioritizing racialized status, even above gender and ethnic culture (see Jones & Shorter-Gooden, 2003). West Indian immigrant women, on the other hand, can *attempt* to distance themselves from the stereotypical images of black (African American) women by highlighting their "West Indian immigrant woman" status in order to be seen more positively by White European Americans (see Model, 2008; Waters, 1999). This status is linked to alternative stereotypes of being family-oriented, hardworking, and compliant (Waters), which, although similar to the mammy stereotype of Black women, starkly contrasts with the jezebel, matriarch, and welfare-queen images. While possibly attenuating negative attention from White European Americans, attempting to mute one's racialized group by prioritizing nativity is likely to be perceived by some African American women, particularly those of the working class, as colluding with the oppressor (Waters), creating an "us vs. them" scenario. Recent media coverage about the success of West Indian immigrants in the US reinforces this divide, simultaneously sending the message that racism is no longer an issue in the Black community (Model).

Black immigrant women may not understand these power dynamics as the basis of the interpersonal divide. For example, the vignette above shows how, I, Roxanne, incorrectly assumed that the relational tensions were from African American women's negative judgments of my nativity and multiracial "look." These were things I had no control over and no responsibility for changing. It was not until I listened carefully to the pain behind the judgments that I began to reassess what the question "Are you black?" might mean when asked by African American women. While the question may seem to be about my racialized identity – do I identify as black, multiracial, or some other race, the deeper meaning is about my black consciousness – do I fully understand, within *and* beyond my personal experience, what it means to be black in the dominant US racialized society where meanings are imposed by White European American men.

US born and immigrant East Asian women also struggle with internalizing or resisting oppressive definitions in a White European American male context (Kim, 2006). East Asian immigrant women may expect that the discrimination they experience from being seen as foreign will decrease as they become more culturally and linguistically proficient in the US context, even as they maintain their Asian cultural affiliations and proficiencies. Given their lack of racialization experience and the East Asian cultural emphasis on collectivism, familial obligation and harmony, hierarchy and conformity (Suh, 2007; Uba, 1994), East Asian immigrant women may not understand why US born East Asian American women seem unable to hold on to their heritage culture while adapting to US culture and language. They may view the attempts of US born East Asian women to deal with racism as personal choices to reject home, culture, and family; such a choice would be viewed as inherently "un-Asian," disrespectful and negative.

In contrast, US born Asian Americans experience White European American assumptions of being foreign *as* racism (e.g. see Liang, Li & Kim, 2004; Sue et al., 2007). They may attempt to resist the stereotypes of Asian American women as forever foreign by being particularly "American," which may mean distancing from being Asian or from home country practices – similar to the "conformity" stage described in racial identity (Kim, 2006). Or they may have more distance from home country practices due to historical oppression or family members' attempts at coping, as was the case in my, Karen's, experience. Regardless of the cause, the resulting lack of shared cultural experiences inevitably distances such a woman from Asian immigrant women who behave in, practice, and value more enculturated ways. More extreme internalization of racist stereotypes by US born Asian women may contribute to reactive denigration of Asian culture and judgments of immigrants as "FOBs" – those Asians that are "too" Asian (Kim, 2006; Pyke & Dang, 2003; Pyke & Johnson, 2003).

As our vignettes illustrate, oppressive intersectional meanings of race, gender and nativity shaped within the white European American male-dominated context to support oppressive hierarchies can be internalized, particularly by those who may position themselves as exceptions to the stereotypes (e.g. West Indian Black immigrant women and East Asian US born women). Taking these "exceptional" positions can create problematic in-group dynamics that have the potential to minimize commonalities and emphasize differences in divisive ways. Black and Asian women then judge each other across nativity differences, oftentimes without knowledge that this essentialized analysis of the other is influenced and constrained by gendered racist stereotypes.[3] Understanding these processes and the pain that immigrant and US born women are experiencing from oppression can help us reach across the imposed divide.

Creating unity to resist oppression

> "Are we really doing this chapter?"
>
> "I don't know, are we?"
>
> "I'm not sure I want to. I'm scared."
>
> "I'm scared too. But doesn't that mean we have to?"
>
> "What if we're honest and we offend someone? What if we aren't good enough and can't see all of our privilege? Someone in the other space might be mad or, worse, hurt."
>
> "What if the pain of remembering is too much? What if we can't get past our own anger? It's different to talk about this between us than to put it out there."
>
> "What if all those things happen and we use our pain as a bridge to something bigger?"

Although we were clearly fearful of writing this chapter, what held us was the promise of this last "What if...". What if we could reach across the divide to have genuine conversations, to foster authentic relationships that would enable us to work together across differences in nativity against racism and sexism within the dominant White European American male context and within the contexts of male-dominated communities of color? What if we could foster such connections in our professional work? The latter question was especially meaningful to us because in our roles as therapists, educators, and researchers, we have power. With that power, we believe, comes an ethical responsibility to deepen the basis of our cultural sensitivity and competence; to use nuanced power analyses to avoid reifying the stereotypes we are attempting to resist; and to build relational coalitions across differences to promote social justice (Toporek & Suyemoto, 2014).

One foundational aspect of reaching across the divide is recognizing that intersectionality is not only about multiple interacting oppressions but also about intersections of privilege and oppression, even if that privilege is relative and ascribed rather than dominant (Cole, 2009). This means that both US born women and immigrant women must be willing to acknowledge and work through their privilege as well as their oppression, as the very basis of developing self-awareness, understanding others' experience, and fostering effective relational processes. This triad is widely acknowledged as the foundation of cultural sensitivity or social justice therapy (American Psychological Association, 2003).

An analysis of structural power is a second foundational aspect. This analysis helps to locate the problem in the oppressive system rather than in the women who are struggling to find strategies to thrive within that system. Our analysis above, for example, indicates that part of the pain felt by immigrant Black and Asian women is perceived to be caused by US born Black and Asian women and vice versa. But the attitudes and actions causing this pain are not deliberate. They are rooted instead in a lack of awareness of how each has relative and ascribed privilege in relation to the other *within a context that oppresses both.* The divide between immigrant and US born women is subtly and insidiously created within gendered racism. Immigrant and US born Black and Asian women are socialized into a negative-sum game where differences serve divisiveness that maintain hierarchies of relative and ascribed power and privilege (Lipsitz, 2001). Recognizing that the ways in which Black and Asian immigrant and US born women hurt each other are rooted in imposed discrimination illuminates the ways in which pain and oppression is shared rather than fostering oppositional positioning.

An analysis of how structural power contributes to divides also enables women of color, across nativity, to more successfully manage the painful experience of working through oppression and privilege. Such an analysis highlights how any given individual is not responsible for injustice or unearned advantages. This can prevent becoming mired in anger and pain from one's own oppression or in guilt and shame from recognizing privilege. These emotions prevent being able to fully validate the pain of another or to fully examine the benefits from personal privilege. However, this analysis cannot be depersonalized to the point of becoming an intellectual exercise, because authentic connection must involve personal investment, affective as well as cognitive understanding and empathy.

Although this awareness and analysis is foundational, reaching across the divide is ultimately an interpersonal risk. Our experience is that reaching across the divide involves a willingness to confront and experience our own pain *while also recognizing and validating the pain of another.* We each brought to this chapter the foundation of the structural power analysis presented here. But it was the personal and interpersonal applications of that analysis that were most challenging. Talking together, we found that experiencing validation of our pain from the "other" across nativity made it easier to let it go, to not have to continually over-prove the reality and centrality of our own oppression. We could then build upon this sense of validation and recognition, to invite guidance and new insights from the other for our own growth. This was not easy.

There were many times when we felt defensive or attacked by each other when blind spots were pointed out. However, in the end, this process was well worth it because through it we learned and grew enormously as individuals and in our relationships. This growth enables us to better contribute to change that we value, to maintain a personal integrity that is central to developing as an ally (Smith & Reddington, 2010) and, most importantly, to maintain our professional ethical responsibility to ensure cultural sensitivity and contribute to social justice.

In sum, reaching across the imposed divide between immigrant Black and Asian women on one hand and US born Black and Asian women on another is similar to creating connections across any major system of power and privilege. Continuous self-examination, active efforts to share power and undermine hierarchy, the sharing of voice and raising awareness of how the personal *is* political (Goodman et al., 2004) are all tools that we can bring from our feminist and multicultural understandings. Our experience is that it is possible to utilize one's own pain as the basis of empathy and relational connection for another by integrating self-awareness, power analysis, and perspective taking that is fostered by a deeper understanding and exploration of another's experiences.

> Although all your cultures reject the idea that you can know the other, you believe that besides love, pain might open this closed passage by reaching through the wound to connect.... Like love, pain might trigger compassion – if you're tender with yourself, you can be tender to others.... [W]hy not use pain as a conduit to recognizing another's suffering, even that of the one who inflicted the pain? (Anzaldúa, 2002, pp. 571–572)

Notes

1. We are using the language "Black" and "Asian" throughout the chapter to denote the racialized groups in the US, and *not* as any indicator or differentiator of nativity.
2. We place "invisible" in quotations because the privilege is invisible to only the one holding it, but it is frequently quite visible to those who lack it. Our thanks to Jazzmine Williams for helping us to articulate this point.
3. We'd like to acknowledge here that these interpersonal dynamics are, of course, influenced by many other things as well, perhaps particularly by the gendered power held by men and enacted in relations with White men and men of color. By focusing here on the ways nativity intersects with gendered racism, we do not negate the influence of racialized or nativity-related sexism. As noted in our introduction, space constrains what is possible to examine in one place and time.

References

Abrams, K. (2005). Polygamy, prostitution, and the federalization of immigration law. *Columbia Law Review, 105*, pp. 641–716.

American Psychological Association (2003). Guidelines on multicultural education, training, research, practice, and organizational change for psychologists. *American Psychologist, 58*, pp. 377–402.

Ancheta, A.N. (2006). *Race, rights, and the Asian American experience* (2nd ed.). New Brunswick, NJ: Rutgers University Press.

Anzaldúa, G.E. (2002). Now let us shift…the path of concocimiento…inner work, public acts. In G.E. Anzaldúa & A. Keating (Eds.), *This bridge we call home: Radical visions for transformation* (pp. 540–578). New York: Routledge.

Benson, J.E. (2006). Exploring the racial identities of Black immigrants in the United States. *Sociological Forum, 21*, pp. 219–247.

Cokley, K.O. (2002). Testing Cross's revised racial identity model: An examination of the relationship between racial identity and internalized racialism. *Journal of Counseling Psychology, 49*(4), pp. 476–483. doi:10.1037/0022-0167.49.4.476.

Cole, E.R. (2009). Intersectionality and research in psychology. *American Psychologist, 64*, pp. 170–180.

Collins, P.H. (2000). *Black feminist thought.* New York: Routledge.

Deaux, K., Reid, A., Martin, D. & Bikmen, N. (2006). Ideologies of diversity and inequality: Predicting collective action in groups varying in ethnicity and immigrant status. *Political Psychology, 27*(1), pp. 123–146. doi:10.1111/j.1467-9221.2006.00462.x.

Devos, T. & Banaji, M.R. (2005). American = White? *Journal of Personality and Social Psychology, 88*(3), pp. 447–466.

Donovan, R.A. (2011). Tough or Tender: (Dis)Similarities in White college students' perceptions of Black and White women. *Psychology of Women Quarterly, 35*, pp. 458–468.

Donovan, R.A. (2007). To blame or not to blame: Influences of target race and observer sex on rape blame attribution. *The Journal of Interpersonal Violence, 22*, pp. 722–736.

Donovan, R. & Williams, M. (2002). Living at the intersection: Effects of racism and sexism on Black rape survivors. *Women and Therapy, 25*, pp. 95–105.

Dovidio, J.F., Gaertner, S.L., Kawakami, K. & Hodson, G. (2002). Why can't we all just get along? Interpersonal biases and interracial distrust. *Cultural Diversity and Mental Health, 8*, pp. 88–102. doi:10.1037//1099-9809.8.2.88.

Espin, O. & Gawelek, M.A. (1992). Women's diversity: Ethnicity, race, class, and gender in theories of feminist psychology. In L. Brown & M. Ballou (Eds.), *Personality and psychopathology: Feminist reappraisals* (pp. 99–108). New York: Guilford Press.

Finlay, L. (2002). "Outing" the researcher: The provenance, process, and practice of reflexivity. *Qualitative Health Research, 12*, pp. 531–545.

Ghavami, N. & Peplau, L.A. (2013). An intersectional analysis of gender and ethnic stereotypes: Testing three hypotheses. *Psychology of Women Quarterly, 37*(1), pp. 113–127.

Goodman, L.A., Liang, B., Helms, J.E., Latta, R.E., Sparks, E. & Weintraub, S.R. (2004). Training counseling psychologists as social justice agents: Feminist and multicultural principles in action. *The Counseling Psychologist, 32*(6), pp. 793–837. doi:10.1177/0011000004268802

Gupta, A., Szymanski, D.M. & Leong, F.L. (2011). The "model minority myth": Internalized racialism of positive stereotypes as correlates of psychological distress, and attitudes toward help-seeking. *Asian American Journal of Psychology, 2*(2), pp. 101–114. doi:10.1037/a0024183

Guthrie, R.V. (2004). *Even the rat was White: A historical view of psychology* (2nd ed.). Boston: Pearson.

Hoeffel, E.F., Rastogi, S., Kim, M.O. & Shahid, H. (2012). *The Asian population: 2010.* United States Census Bureau. Retrieved from http://www.census.gov/prod/cen2010/briefs/c2010br-11.pdf

Hong, G.K. & Ham, M.D.C. (2001). *Psychotherapy and counseling with Asian American clients: A practical guide.* Thousand Oaks, CA: Sage.

Jones, C. & Shorter-Gooden, K. (2003). *Shifting: The double lives of Black women in America.* New York: Perennial.

Kim, H. (2006). *Asianized Asians, Twinkies, and North Face puffy jackets: Constructing racialized gender identities among second generation Korean American college women.* Paper presented at the American Sociological Association Annual Meeting, Montreal, Canada. *SocINDEX with Full Text,* EBSCO*host* (accessed May 29, 2013).

Kinsley, C.H. (1994). Questions people have asked me. Questions I have asked myself. In C. Camper (Ed.), *Miscegenation blues: Voices of mixed race women* (pp. 113–132). Toronto, Canada: Sister Vision.

Liang, C.T.H., Li, L.C. & Kim, B.K.S. (2004). The Asian American racism-related stress inventory: Development, factor analysis, reliability, and validity. *Journal of Counseling Psychology, 51*, pp. 103–114.

Lipsitz, G. (2001). *American studies in a moment of danger.* Minneapolis, MN: University of Minnesota Press.

McIntosh, P. (1997). White privilege: Unpacking the invisible knapsack. In V. Cyrus (Ed.), *Experiencing race, class, and gender in the United States* (2nd ed.) (pp. 194–197). Mountain View, CA: Mayfield.

Model, S. (2008). The secret of West Indian success. *Global Society, 45*, pp. 544–548.

Moradi, B. & DeBlaere, C. (2010). Women's experiences of sexist discrimination: Review of research and directions for centralizing race, ethnicity and culture. In H. Landrine & N.F. Russo (Eds.), *Handbook of Diversity in Feminist Psychology.* New York: Springer Publications.

Park, J.Z. (2008). Second-generation Asian American pan-ethnic identity: Pluralized meanings of a racial label. *Sociological Perspectives, 51*(3), pp. 541–561.

Patton, T.O. (2001). "Ally McBeal" and her homies: The reification of White stereotypes of the other. *Journal of Black Studies, 32*, pp. 229–260.

Prasso, S. (2006). *The Asian Mystique: Dragon ladies, geisha girls, and our fantasies of the exotic Orient.* New York: PublicAffairs.

Purdie-Vaughns, V. & Eibach, R.P. (2008). Intersectional invisibility: The distinctive advantages and disadvantages of multiple subordinate group identities. *Sex Roles, 59*, pp. 377–391.

Pyke, K.D. & Dang, T. (2003). 'FOB' and 'Whitewashed': Identity and internalized racism among second generation Asian Americans. *Qualitative Sociology, 26*, pp. 147–172.

Pyke, K.D. & Johnson, D.L. (2003). Asian American women and racialized femininities: "Doing" gender across cultural worlds. *Gender & Society, 17*(1), pp. 33–53. doi:10.1177/0891243202238977

Romero, R.E. (2000). The icon of the strong Black woman: The paradox of strength. In L.C. Jackson & B. Greene (Eds.), *Psychotherapy with African American women: Innovation in psychodynamic perspectives and practice* (pp. 225–238). New York: Guilford Press.

Rong, X.L. & Fitchett, P. (2008). Socialization and identity transformation of Black immigrant youth in the United States. *Theory into Practice, 47*, pp. 35–42.

Schwartz, S.J., Unger, J.B, Zamboanga, B.L. & Szapocznik, J. (2010). Rethinking the concept of acculturation: Implications for theory and research. *American Psychologist, 65*, pp. 237–251.

Smedley, A. & Smedley, B.D. (2005). Race as biology is fiction, racism as a social problem is real: Anthropological and historical perspectives on the social construction of race. *American Psychologist, 60*, pp. 16–26.

Smith, L. & Reddington, R.M. (2010). Lessons from the experiences of White anti-racist activists. *Professional Psychology, 41*, pp. 541–549.

Steele, C.M., Spencer, S.J. & Aronson, J. (2012). Contending with group image: The psychology of stereotype and social identity threat. *Advances in Experimental Social Psychology, 34*, pp. 379–440.

Sue, D.W., Capodilupo, C.M., Torino, G.C., Bucceri, J.M., Holder, A.M.B., Nadal, K.L. & Esquilin, M. (2007). Racial microaggressions in everyday life: Implications for clinical practice. *American Psychologist, 62*, pp. 271–286.

Suh, S.H. (2007). Too maternal and not womanly enough: Asian-American women's gender identity conflict. *Women & Therapy, 30*(3–4), pp. 35–50.

Suro, R., Kocchar, R., Passel, J., Escobar G., Tafoya, S., Fry, R. & Wunsch, M. (2007). Asians: 2004. *American Community Survey Reports*. U.S. Census Bureau. Retrieved from http://www.census.gov/prod/2007pubs/acs-05.pdf

Suyemoto, K.L. (2002). Redefining "Asian American" identity: Reflections on differentiating ethnic and racial identities for Asian American individuals and communities. In L. Zhan (Ed.), *Asian Americans: Vulnerable populations, model interventions, and clarifying agendas* (pp. 195–231). Boston: Jones and Bartlett.

Thomas, K.J.A. (2012). *A demographic profile of Black Caribbean immigrants in the United States*. Washington, DC: Migration Policy Institute.

Toporek, R.L. & Suyemoto, K.L. (2014). Social justice in counseling and clinical psychology. In C.V. Johnson, H.L. Friedman, J. Diaz, Z. Franco & B.K. Nastasi (Eds.), *Handbook of social justice and psychology, Volume 3: Youth and disciplines in psychology*. Santa Barbara, CA: Praeger.

Uba, L. (1994). *Asian Americans: Personality patterns, identity, and mental health*. New York: The Guilford Press.

U.S. Census Bureau (2011). *The Black alone on in combination population in the United States: 2011*. Retrieved from http://www.census.gov/population/race/data/ppl-bc11.html.

Vasquez, H. & McGraw, S. (2005). Building relationships across privilege: Becoming an ally in the therapeutic relationship. In M. Mirkin, K.L. Suyemoto & B. Okun (Eds.), *Psychotherapy with women: Exploring diverse contexts and identities*. New York: Guilford Press.

Waters, M.C. (1999). *Black identities: West Indian immigrant dreams and American realities*. New York: Russell Sage Foundation.

White, D.G. (1985). *Ar'n't I a woman? Female slaves in the plantation south*. New York: Norton.

Part II

Work, Social Class and "Traditional" Gender Roles

Introduction to Part II

Oliva M. Espín and Andrea L. Dottolo

Work is the central motivation for migration for many women. In some cases, it is their primary reason to migrate. In other cases, it is the reason they use to justify the decision to move to another country. In yet other situations, the need to work to support themselves and their families becomes urgent upon arrival in the new context.

This section focuses on issues of work confronted by many immigrant women, including balancing those issues with mothering responsibilities. Even though other chapters in this book mention issues of work, these chapters emphasize the possibilities and stresses created by employment for immigrant women.

Finding remunerated work in the host society is central to the financial stability and emotional adaptation of immigrants. But finding employment may present a special challenge because of language difficulties or other factors. Although many women migrate primarily to better their financial situation, some experience a dramatic downturn in terms of their jobs and careers. Contrary to stereotypes, many women immigrating to the US have higher levels of educational than the national averages of their home countries, and they often desire to return home. As with all immigrants, they may face unemployment, underemployment and disqualification of their previously held professional credentials, including a loss of professional status in moving from being a scientist, a teacher or a clerical worker to a domestic worker (Berger, 2004). Moreover, immigrant women are disproportionately employed in physically demanding jobs and in industries where there is substantial potential for injury. Earnings for immigrant women are typically low (Larsen, 2004), particularly in the case of undocumented women immigrants who have no legal rights. They may suffer emotional (e.g. harassment), physical (e.g. poor working conditions) and economic (e.g. no consistent pay or health benefits) abuse by US employers.

Despite these challenges and difficulties, finding work is a mechanism of both material stability and psychological well-being for women immigrants (Yakushko & Espín, 2010). Aside from the financial reliability provided by a job, becoming wage earners may give women an increased status either in the new context or because of the importance of their remittances for the family back home.

One of the typical patterns of migration is the "mother first" relocation in which women enter the US through their employment status as domestic workers, leaving their children with relatives until they can be reunited (Berger, 2004). There are innumerable studies (e.g. Cortina & Ochoa-Reza, 2013) dedicated to exploring the impact of remittances by migrants of both sexes on the economies of their home countries. But many women who work to send money home do not have an easy time. Immigrant women separated from their children may be more likely to suffer from depression than those women who either were not separated from their children or were not mothers. Separation from children and extended families creates loss of social support and feelings of isolation. This situation has been associated with poor mental health among women who are residing in the US both with and without legal documentation (Berger, 2004).

In keeping with the focus of this volume, we cannot ignore "the impact of immigrant women's regular wage work has on gender relations" (Pessar 2003, p. 27). Paid employment may help them claim some power in the family structure, including in their marriages. Despite their frequently disadvantaged position in US society, "[e]mployed immigrant women generally gain greater personal autonomy and independence.... [W]omen's regular access to wages and their greater contribution to household sustenance frequently lead to more control over budgeting and other realms of domestic decision making (Pessar 2003, p. 27). Indeed, there is evidence that "migration and settlement bring changes in traditional patriarchal arrangements" (Pessar 2003, p. 27). Many women learn to contest the patriarchal narratives of ethnic solidarity and thus change the content of some of those narratives (Pessar, 2003).

However, many women do not achieve much for themselves, because "many women work long hours, often unpaid, in family businesses" (Kofman et al., 2000, p. 41), significantly contributing to the financial success of the household. Indeed, in many cases, "the success of small-businesses [headed by immigrants] comes at the expense of subordinated immigrant women" (Pessar, 2003, p. 24). "We need to look more closely at the gender relations that underpin so many of these enterprises [because] social class categories that define women's position according to that of the (male) head of household are always unsatisfactory, but in this instance are virtually meaningless" (p. 41).

Furthermore, while this section focuses mostly on work and employment, this topic is situated within a larger framework of social class where finances are only one aspect. Immigrant women must also contend with navigating foreign rules of cultural capital, which include all the other markers of class, including prestige, education, taste and "identity management" (Zweigenhaft & Domhoff, 1998). Even in the rare cases where immigrants are financially successful in the host country, they are still often read as being of lower class status due to their "foreignness". In the US, middle class standards of dress, emotional expression (or lack thereof), individualism, language and specific ideologies of gender, race and sexuality must be adhered to in order to become a member of the "club" of the middle-upper class.

In this section, Ma and Shea explore the literature on first-generation East Asian immigrant women's experiences with employment and its effects on "traditional" gender roles and mothering. Their discussion is grounded in a feminist therapy framework. They review the literature on first-generation East Asian immigrant women's experiences with employment and the changing gender-role dynamics in the context of migration. Specifically, they highlight the impact of cultural ideologies, ethnicity, acculturation and social class on East Asian immigrant women's work decision, mothering experience as well as psychological health. Empowering and disempowering aspects of their employment and changing gender roles are discussed. Suggestions for clinical interventions with this population are provided.

Balan and Mahalingam present an empirical investigation of intersecting immigrant cultural expectations and gendered processes as they affect South Asian immigrant working mothers in their struggles to be "good Asian moms". Through detailed conversations, Ahmed-Ghosh discusses experiences of different generations of Afghan refugee women. She deconstructs the concept of "home" to discuss their experiences of migration to the USA. Ahmed-Ghosh elaborates on their experiences as mothers in the diaspora – specifically, the gendered differences based on the wave of migration from Afghanistan, class situation in the USA, and experiences based on the US-Afghan conflict. Even though Ahmed-Ghosh's chapter does not focus on issues of work per se, they are present in the experiences of the Afghan women she interviewed. The effects of social class on the experiences of the different groups of women portrayed in this section, even though not explicitly discussed in all chapters, are present in their understandings of both work and mothering.

We believe that deeper investigation of these issues is needed in the future. We do hope that by raising these questions we will encourage others to further pursue the experiences of immigrant women in relation to work.

References

Berger, R. (2004). *Immigrant women tell their stories.* New York: The Haworth Press.

Cortina, J. & Ochoa-Reza, E. (Eds.) (2013a). *New perspectives on international migration and development.* New York: Columbia University Press.

Cortina, J. & Ochoa-Reza, E. (2013b). Women, children, and migration. In J. Cortina & E. Ochoa-Reza (Eds.), *New perspectives on international migration and development* (pp. 139–163). New York: Columbia University Press.

Kofman, E., Phizacklea, A., Raghuram, P. & Sales, R. (2000). *Gender and international migration in Europe: Employment, welfare and politics.* New York: Routledge.

Larsen, L.J. (2004). The foreign-born population in the United States: 2003. *Current population reports.* Washington, DC: U.S. Census Bureau.

Pessar, P.R. (2003). Engendering migration studies. In P. Hondagneu-Sotelo (Ed.), *Gender and U.S. migration* (pp. 20–42). Berkeley, CA: University of California Press.

Yakushko, O. & Espín, O.M. (2010). The experience of immigrant and refugee women: Psychological issues. In H. Landrine & N.F. Russo (Eds.), *Handbook of diversity in feminist psychology* (pp. 535–558). New York: Springer.

Zweigenhaft, R.L. & Domhoff, G.W. (1998, March/April). The new power elite. *Mother Jones, 23*(2), pp. 44–47.

3

Work and Gender Roles among East Asian Immigrant Women in the United States

Pei-Wen Winnie Ma and Munyi Shea

Asians are the second fastest growing ethnic minority group in the United States (U.S. Census Bureau, 2012a) and are projected to more than double between 2012 and 2060 – growing to 34.4 million and making up 8 percent of the total population (U.S. Census Bureau, 2011). In a report released by the Pew Research Center, Asians accounted for 25 percent of all the first-generation immigrants in 2012, of which 54 percent are women (Cohn, 2013). The majority of the recent immigrants come from the People's Republic of China (or referred as mainland China) and reside in urban cities, including Los Angeles, California and New York City (McCabe, 2012), where their children attend urban public schools (Fix & Capps, 2005). In spite of their rapid population growth, psychological research on Asian immigrants – especially those of recent immigration status and of women – remains scant (U.S. Department of Health and Human Services, 2001). Within the limited literature relevant to this population, there are even fewer studies exploring East Asian immigrant women's experience of employment and gender roles, two aspects that are salient to their identity development and mental health.

The purpose of this chapter is to draw upon existing literature in psychology, sociology, women's studies and ethnographic reports to illuminate the experience of first-generation Asian immigrant women, with a focus on their employment, motherhood, psychological health and related clinical implications. A few notes should be made about the premises of our discussion: First, the Asian population is remarkably diverse in terms of its ethnic groups, cultures, languages, religions and philosophies (Chung & Bemak, 2007). The unique set of beliefs and values shaped by the mythology and religion among South Asian Indians (Dasgupta, 1998) is likely to give rise to specific worldviews and behavioral patterns that are rather different from those of East Asians.

Thus we focused our discussion primarily on East Asian (China, Japan, Hong Kong, Korea, and Taiwan) women, with limited description of Vietnamese women and Filipina. Second, our literature review focused more on the experience of Asian immigrant wives and mothers, as the shifting of gender roles during the immigration process has a tremendous impact on their vocational and domestic lives, as well as psychological health. Lastly, we would like to acknowledge that immigrant women from various ethnic groups share similar challenges with regard to their vocational life (Yakushko, 2006). Hence, our goal is to highlight some of the unique experiences within East Asian immigrant groups, but we will also incorporate literature that pertains to immigrant women in North America as a whole.

Theoretical framework

In this chapter, we will critically analyze the literature from a feminist therapy, especially a feminist family therapy framework (e.g. Hare-Mustin, 1978; McGoldrick, Anderson & Walsh, 1989; Silverstein & Goodrich, 2003). This framework was chosen because it aligns with both authors' training background (multicultural counseling psychology), research emphasis (i.e. social constructionist perspective) and/or clinical practice with children and families. It would be impossible for us to pinpoint a single feminist theory that guided our discussion or to provide an overview of a variety of feminist theories and stances within the scope of this chapter (see Enns, 2004 for a review on feminist theory and psychotherapy). Our analysis is based upon several important tenets of feminist family therapy, which include 1) a belief that the power of patriarchy permeates all social contexts and interpersonal encounters, including the life of families, therapy and society, and that this power creates and maintains unequal relations between men and women as well as between the dominant racial-cultural group and minority groups (e.g. Liamputtong, 2001); 2) a willingness to question the traditional practice of gender socialization, to challenge the dominant discourse of gender role expectations and to examine the impact of internalized stereotyped sex roles and ideologies of femininity on family lives (e.g. Hays, 1996; Lupton, 2000; Moon, 2003); 3) a consideration of the influences of race, ethnicity, class, religion and other important aspects of women's identities in the analysis of power and privilege (Silverstein & Goodrich, 2003); and 4) a conscious choice to empower women by bringing their experiences and stories – such as sexism, discrimination, oppression, family violence and poverty – to the foreground, making gender issues central to the dialogues of research, assessment and therapy (Brown, 2010; Hare-Mustin, 1978; McGoldrick et al., 1989).

Asian immigrant women's employment in the context of migration

Employment is a salient theme in immigrants' lives in the new society. Throughout the history of the US, adult immigrants came to this land with the hope of providing better educational or financial opportunities for themselves and for their families. Immigrant women are an especially important part of this work force. According to the latest census data (Camarota, 2012), nearly 70 percent of the immigrant women in the US between 25 and 55 are in the work force (either looking for a job or holding a job). Compared to a century ago when mostly immigrant young adult daughters were the female wage-earners of the household, today, immigrant wives and mothers are taking on this role (Foner, 2005).

Although Asian immigrants as a group are more likely than the overall immigrant population to have post-secondary education and higher-level occupations, there is a great deal of within group variation (Batalova, 2011). Asian immigrant women are currently situated in both the higher and lower end of the social strata. Those women who came to the US with a high school or above education and those who pursued advanced education in the US are likely to work in professional fields, such as management, business, finance, education, nursing and administrative support (Batalova, 2011; U.S. Census Bureau, 2012b). On the other hand, working-class Asian immigrant women tend to be heavily represented in operative (e.g. garment factory), clerical and private household occupations (e.g. as nannies or maids) (Loo, 1998).

Asian immigrant women's motivation to work may also vary as a function of their educational attainment and social class. Highly educated professional women, for example, typically come to the US to obtain an advanced degree or to seek better employment opportunities. They are more likely to pursue a career outside of home for psychological fulfillment and financial reward (Blair-Loy, 2003; Greenman, 2011). In some cases, Asian immigrant women are the primary immigrants who obtain legal status to work in the US and bring their spouse as a dependent (Espiritu, 1999; Greenman, 2011). For working-class immigrant women, however, employment outside of the home is often not based on personal interests or aspirations; rather, it is based on financial necessity. This is especially true for those who enter the US with low educational attainment and skills (Espiritu, 1999). Similarly, undocumented immigrants that come to the US through human trafficking also have to take on any available jobs to pay off their debts to the "snakeheads" (human smugglers) (Chin, 1999).

Downward social mobility (e.g. Kim, Conway-Turner, Sherif-Trask & Woolfolk, 2006; Qin, 2009) is another major influence on Asian immigrant women's decision to work outside of the home. In spite of their previous job credentials, many middle-class immigrants, especially men, are not able to secure the same occupational status after their migration, due to language barriers, lack of professional network and systemic racism and discrimination (Qin, 2009). A phenomenon worth noting is that in some labor-intensive industries employers actually prefer to hire immigrant women over immigrant men due to the assumption that these women "can afford to work for less, do not mind dead-end jobs, and are more suited physiologically to certain kinds of detailed and routine work" (Espiritu, 1999, p. 639). As a result, immigrant women who previously stayed at home may decide to look for paid employment out of financial necessity and family well-being (Grahame, 2003; Kim et al., 2006; Qin, 2009). These women's decisions contrast with the assumptions of major career development theories, the latter of which are more individualistically oriented and emphasize individuals' autonomy to choose careers based on their self-concept and personality (Holland, 1973; Super, 1957). A culturally and contextually relevant career framework should seek to examine Asian immigrant women's work decisions within the interlocking systems of immigration and acculturation, gender and social class.

Impact of employment on family dynamics

The elevated role of Asian immigrant women as a co-provider or a primary provider for the family transforms the patriarchal family relationships and women's status in the family.

For Asian immigrant women with a professional career, they are less likely to adhere to transitional gender roles post-migration (Greenman, 2011). An equitable division of labor is more likely to be found in their households. Due to the similar earning power between professional men and women, the husbands are more likely to take on housework compared to their counterparts in the working-class families. However, it has been reported that these women have to use a conflict and confrontation strategy to alter the distribution of labor. Otherwise, they still have to juggle full-time work, housework and childcare responsibilities (Espiritu, 1999).

For East Asian immigrant women who adhere to traditional gender role expectations (e.g. a woman's primary role is to stay at home) prior to migration, their employment outside the home post-migration could be viewed as a source of empowerment, particularly in the forms of independence and the opportunity to challenge patriarchal norms within their families (e.g. Louie, 2004; Rolls & Chamberlain, 2004). It was reported that Vietnamese immigrant women in Philadelphia contributed more financially to household income because their husbands were forced to take on low-wage jobs or stay unemployed post-migration. As a result, these women gained informal family power and would unite together socially to fight against a husband's abuse of authority (Kibria ,1993; Louie, 2004). Zhou and Nordquist (1994) described that paid employment outside of home also provided a "confidence and self-fulfillment which they may never have experienced in traditional Chinese society" (p. 201). Lee and Mock (2005), as cited in Foner (2005), also identified that in contemporary Chinese immigrant families, a traditional patriarchal family is transformed into a system where a mother shares decision-making power with the father.

The process of empowerment associated with employment, however, is frequently accompanied by a simultaneous process of disempowerment in immigrant women's lives (Park, 2008). For example, the working-class Chinese immigrant women in New York City's Chinatown – despite working long hours under adverse conditions – are still expected by their husbands to be solely responsible for childcare and household chores (Loo, 1998; Foner, 1998). These are the "double burden" both professional and working-class immigrant women seem to bear as a result of paid employment. Patriarchal structure in the household remains largely unchanged, and women are still locked into a subordinate position (Foner, 1998; Foner, 2005; Park, 2008).

East Asian immigrant women with paid employment may also notice changes in their relationships with their children. Not only do they have less time to spend with their children, they must also delegate some of the household chores to the "little adults" at home (Kim et al., 2006). Older children often step up and take on additional responsibilities (e.g. baby-sitting younger siblings and cooking) while their mothers work outside of the home (Grahame, 2003; Qin, 2009). The role reversal is also evident on a psychological level, where children share the burden and worries of their family by counseling their stressed mothers, which may potentially add challenges to these children's own cultural adjustment (Qin, 2009).

Impact of employment on psychological health

As previously noted, first-generation East Asian immigrant women tend to face the double burden of household responsibilities and high demands from their work. However, there are limited studies exploring the impact of role overload on East Asian immigrant women's mental and psychological health. In one study Noh and colleagues (1992) found that Korean immigrant women employed outside of home were eight times more likely to exhibit depressive syndromes than employed Korean immigrant men. And surprisingly, the gender difference in depression was most remarkable between the highly educated professional women and their male counterparts. One possible explanation was that working a demanding job outside home made household and parenting responsibilities even more difficult and challenging for the professional women because their family still observed traditional gender roles and gendered division of labor. This result not only supports the double burden hypothesis but also speaks to a larger issue of how employment outside of home differentially impacts working and middle-class immigrant women and men. For one, immigrant women often find the negotiation of their roles and responsibilities in the family stressful and difficult. As several researchers have noted (e.g. Foner, 2005; Grahame, 2003; Qin, 2009), the gendered division of labor is difficult to change in spite of the immigrant women's elevated status through financial independence and contribution to family income. Deeply entrenched gender ideologies and norms in the Asian community may perpetuate gender inequalities and contribute to Asian immigrant women's dysphoria.

Working-class East Asian immigrant women may also experience heightened stress at home due to the changing gender roles and rules when their husbands are underemployment or unemployment post-migration. Min (2001) noted the "status anxiety" Korean husbands would experience due to gender-role reversal and the significant increase of marital conflict among couples. Kim and Ryu (2005) reported that domestic violence is likely to occur in Korean American families when a husband is unemployed and feels marginalized by his limited power at home and in society. Subsequently, the husband would assert his dominance and control over his wife by engaging in violence.

Work and mothering intertwined

An important aspect of East Asian immigrant women's identity and femininity is their role as a mother. Socialized within the patriarchal structure of the society, women were encouraged to center their identity on motherhood (Hochschild, 2002). Religious and traditional cultural beliefs, such as Confucianism, further reinforce the gender norms and behaviors (Grahame, 2003; Qin, 2009). While motherhood generally refers to the qualities of a mother or the state of being a mother, mothering describes the care work performed primarily by women on a physical and emotional level, such as tending to domestic chores, ensuring a comfortable home, nurturing their children and investing in their children's formal and informal education. Instead of assuming the universality of women's motherhood, we posit that mothering is a culture- and context-dependent practice. In the following, we discuss the impact of post-migration employment on Asian immigrant women's decision related to mothering and work and the meaning they attach to mothering. Furthermore, we explore the intersection of ethnicity, acculturation, and social class in shaping immigrant women's mothering practice.

Impact of employment on the meaning of mothering – what is a "good mother"?

Studies suggest that work and motherhood are often inseparable dimensions for ethnic minority and immigrant women; paid employment is an extension of their mothering experience to ensure financial stability and to provide for their children (Collins, 1994; Kim & Kim 1998; Park, 2008). Park (2008) suggested that constructing meaning of their paid employment is a part of immigrant women's cultural adaptation strategy.

In a qualitative interview study (Park, 2008), Korean immigrant women described that their paid employment outside of home was a choice based on economic necessity rather than personal interest. Initially, these women took on low-status jobs in the US and did not find their menial work particularly empowering; to the contrary, they experienced a demotion from their status as middle-class stay-at-home mothers. Overtime, however, they shifted their view toward their paid employment from a stance of reluctance and disempowerment to a stance of acceptance and positive outlook. This change of perspective was brought on by the realization that their own sacrifices would contribute to the upward mobility of their children and would help the family achieve the "American Dream". Thus they were redefining the meaning of their jobs and the meaning of an "ideal mother".

The negotiation of competing priorities at work and at home is often stressful and difficult. Asian immigrant women are at times faced with the decision to continue with paid employment or to retreat into full-time mothering. This decision-making process may be further complicated by contextual factors such as cultural ideologies, immigration, and socioeconomic class. For example, Greenman (2011) analyzed the employment and earning data in the science and engineering fields and concluded that Asian American women have a significantly higher rate of labor force participation when compared to their white female counterparts. Specifically, Asian American women are less likely to take time out of the workforce or have work-hour reduction in response to parenthood.

Several important factors are worth pointing out here. First, the data presented in Geenman's (2011) study highlight the differences in the belief of what constitutes a "good mother". While white women are more likely to prioritize spending time with their children over making more money, East Asian immigrant women, on the other hand, are more likely to regard one's paid employment as a strategy to ensure financial security and to provide their children with a higher quality of life (Park, 2008). Second, while East Asian immigrant men and women adhere to traditional gender roles in many respects, East Asian immigrant men also tend to have higher expectations for women to work and contribute to family income. This phenomenon is common in both working- and middle-class Asian immigrant families, in part due to the immigrants' "survival" mentality (Greenman, 2011; Louie, 2004). Third, reliance on extended family members' support in childcare is a common cultural practice in East Asian societies. Grandparents are often invested in caring for their grandchildren so that women can stay in full-time paid employment (Greenman, 2011). Moon (2003) noted, based on her qualitative interview of 14 middle-class Korean immigrant women, that one distinct pattern of mothering after migration is shared mothering – where women share mothering responsibilities with other female relatives (i.e. their own mothers and mothers-in-law) and husbands. Additionally, Korean immigrant mothers who work outside of the home may rely on Korean churches as a cultural resource for monitoring and socializing their children (Kim et al., 2006).

On the other end of the spectrum, however, there are immigrant women who decide to retreat to full-time mothering after a period of time of struggling to balance work and family responsibilities. This is more common in families where the financial situation has stabilized or where the husbands are not willing to share parenting and household duties (Moon, 2003). Finally, some women endure what Moon (2003) called "isolated and privatized mothering" – where women raise children on their own without any help from family, friends or husbands. It is important to note that not all immigrant families have supportive extended family systems in the US. Immigration to a new country often brings a loss of support systems. Hence, isolated and privatized mothering may be a pragmatic rather than ideological choice.

At times, being a "good mother" also requires Asian immigrant women to physically separate from their children. Due to poverty and lack of employment opportunities in their home country, women from countries such as Indonesia or the Philippines migrate to North America for employment, and they leave their children to be cared for by extended family members back home (Hochschild, 2002). The irony is that their employers in North America are women who hire them as nannies and domestic workers so that these employers can focus on their demanding careers. Needless to say, the lives of these Asian immigrant women who have left their children behind is filled with loneliness and sadness. Interestingly, their coping strategy is to focus on transferring their love to their employers' children (Hochschild, 2002).

A similar but different version of mother-and-child separation is also notable among working-class Chinese immigrants. In order to retain demanding garment factory jobs in the US, some Chinese immigrant women send their newborn infants back to China (Sengupta, 1999). It is estimated that 10–20 percent of the babies from immigrant Chinese families born each year have been sent back to China (Sengupta, 1999), and over 50 percent of expectant Chinese immigrant women in New York City intend to do the same (Kwong Chung, Sun, Chou & Taylor-Shih, 2009). During the separation period, children are cared for by their grandparents or extended families in China, and they return to the US to be reunited with their parents around preschool or elementary school age. This phenomenon has been termed *reverse migration separation* (Kwong et al., 2009) or described as the experiences of *satellite babies in transnational situations* (Bohr & Tse, 2009).

For Chinese immigrant women, the choice to separate from their child is emotional and painful (Kwong et al., 2009). In a qualitative study on the satellite babies phenomenon in New York City (Ma, Hsu, Desai & Zehner, 2011), Chinese immigrant mothers reported that their low wages made it impossible to keep the family alive with only one income. Additionally, long work hours and lack of affordable childcare options in the U.S.US are clearly further constrains. Hence, they had to send their child(ren) back to China so that both parents can focus on working to save up money or pay off the large debt to their smugglers as soon as possible. All of the participants articulated that if they have the financial resources, they would certainly prefer to have their children stay with them.

At present, studies on parent-child migration separation, across various ethnic immigrant groups, have focused primarily on the account of the mothers (e.g. Bohr & Tse, 2009; Hochschild, 2002), or the impact on their children (e.g. Smith, Lalonde, Johnson, 2004; Suarez-Orozco, Todorova, Louie, 2002). The literature on the role of fathers in the decision-making process has been absent. We speculate that since Asian fathers have traditionally taking on a strong breadwinner role and less of a caretaker role, their voices on the issue of childcare arrangement have not been fairly portrayed or represented. It is worthwhile to note that an emerging body of literature started to examine the role of East Asian immigrant fathers as caretakers. It was found that maternal employment outside of home and renegotiation of gender roles post- migration prompted more East Asian fathers to take on more childcare, household chores or even being identified by their children as the primary caretakers (Kanatsu & Chao, 2008).

In sum, the studies reviewed above underscore the importance of understanding Asian immigrant women's work and mothering choices from a cultural and contextual perspective. Specifically, studies by Hochschild (2002) and Ma et al. (2011) illustrate how poverty and lack of access to resources can act as a form of systematic oppression in the lives of working-class Asian immigrant women. Furthermore, Asian immigrant women who have been taught to prioritize the needs of their children and family above their own may internalize guilt and shame about their employment outside of home. They tend to experience feelings of inadequacy as a mother when they cannot spend enough time with their children (Moon, 2003; Qin, 2009). These feelings may exacerbate their risk of developing depression and other psychological symptoms.

Impact of acculturation on mothering practice

Acculturation – the process of adapting to the behaviors, language, values, attitudes, customs and cultural practices of the host country – is likely to influence immigrant parents' expectations regarding their children's development, child-rearing practices and parent-child interactions (Garcia-Coll & Marks, 2009; Kim, Atkinson & Umemoto, 2001). For example, some parents encourage their children to speak only English or to quickly assimilate to the host culture as a way to promote their children's successful adaptation and competitive advantage in the society (Henderson, 2003), whereas other parents emphasize the importance of maintaining one's cultural heritage and mother tongues due to their own limitation in English proficiency or their desire of keeping their children's ethnic identity (Chao & Tseng, 2002; Chang & Karl Kwan, 2009).

Tiger mothers: Truth or myth?

The memoir *Battle Hymn of the Tiger Mother* by Yale Law Professor Amy Chua (Chua, 2011) generated many sensational discussions and controversies about *tiger parenting*, which was said to be commonly practiced in Chinese families. Chua describes tiger parenting as highly intense, strict, critical, demanding and achievement-oriented, which aims to foster children's academic success and a sense of family responsibility. For instance, Chua would constantly remind her daughters that they were falling short of her standard and getting a grade other than A in school was unacceptable. She would make her daughter practice violin for hours – with no bathroom breaks – until her daughter mastered the piece. And she would call her daughters "garbage" if they behaved disrespectfully – something Chua had been called by her critical Chinese parents when she was a young child.

In spite of the book's commercial success, Chua's view of Chinese parenting was deemed "simplistic" and "stereotypical". Recent empirical studies on Chinese families have uncovered more nuances in immigrant Chinese parents' parenting style. In a longitudinal study of 444 first-generation Chinese immigrant families, Kim and colleagues (2013) found that Chinese immigrant parents actually used more *supportive parenting* (high warmth, high positive control) than *tiger parenting* (high warmth, high hostility and negative control) in disciplining their adolescent children. Moreover, supportive parenting rather than tiger parenting was associated with the best child developmental outcomes, including higher academic achievement and lower depression.

Cheah, Leung and Zhou (2013) reported in their qualitative study that although first-generation immigrant Chinese mothers identify with the strictness and harshness of traditional Chinese parenting, they reject the overly critical and controlling aspects of tiger parenting. The authors explained that as these mothers become more acculturated to European American culture, they are also more likely to adopt a balanced view and aspire to incorporate both European American and traditional Chinese values and ideologies in their parenting practices. For instance, these Chinese mothers may want to foster a sense of autonomy in their children while at the same time teach them the importance of maintaining interpersonal relatedness and harmony. These findings, in contrast to Chua's descriptions, suggest that immigrant parents' child-rearing beliefs and practices are fluid and adaptable to the contextual demands.

Native and immigrant mothers' emotion socialization strategies

The interaction between the changing cultural contexts and mothering practice is complex and dynamic. In a recent study on preschool children's emotion socialization, Wang (2012) found that first-generation immigrant Chinese mothers in the US – though immersed in the mainstream American culture that value assertiveness, individuality and emotional expression – are less likely than native Chinese mothers in China to help their toddler children express or understand emotions during a memory sharing and storytelling task. The same author reported similar findings in a prior study where native Chinese mothers, when compared with immigrant Chinese mothers in the US, tend to adopt the narrative and conversation styles of Euro-American mothers (Wang, 2006). Interestingly, regardless of their ethnic and acculturation background, both native Chinese mothers and immigrant Chinese mothers express feelings more frequently with their daughters than with their sons. Reciprocally, daughters express more feelings than sons in their memory sharing with their mothers, which provides evidence for gendered emotional expression and socialization strategies.

Taken together, the aforementioned studies have several developmental and research implications. First, both general parenting orientation (e.g. tiger parenting) and specific emotion socialization strategies have significant impacts on children's developmental outcomes, including self-esteem and emotion regulatory skills, all of which contribute to children's positive feeling states (Wang, 2012). Second, although immigrant mothers tend to adapt to Western ideologies and parenting practices over time (Cheah et al., 2013), they may not become fully acculturated or more acculturated than their native counterparts (Wang, 2012). These findings warrant a more nuanced investigation of immigrant mothers' cultural adaptation processes, including the dimension of enculturation – a process by which immigrants retain or socialize to the values, ideologies, practices and norms of one's indigenous culture. It is possible that the immigrant Chinese mothers in Wang's study were not necessarily less acculturated, but rather, they were more enculturated than the native Chinese mothers. Previous research suggested that adherence to one's indigenous values and ethnic identity may help Asian individuals draw upon cultural resources, such as collectivist style of coping, to counter experiences of racism and discrimination (Yip, Gee & Takeuchi,

2008). Thus, immigrant Chinese mothers may be intentional about retaining their traditional values and helping their children develop a sense of ethnic identity and belonging.

Although gender role socialization was not explicitly discussed in these studies, it was prevalent in these Chinese families. Gender role socialization, like other learning processes, relies on modeling and reinforcement (Addis & Cohane, 2005) whereby "proper" or gender-role-appropriate behavior is rewarded and gender role inappropriate behavior is sanctioned (Bronstein, 2006). For example, the native Chinese mothers and first-generation immigrant Chinese mothers in Wang (2012)'s study seem to have internalized socially acceptable feminine or masculine traits and behaviors (e.g. girls are more sensitive and emotionally expressive than boys); they continue to reinforce these gendered behaviors in their children's lives through specific emotion socialization strategies. Similarly, the shifting of maternal and paternal roles across developmental stages, with fathers becoming closer to tiger parents to older adolescents (Kim et al., 2013), underscore a gendered division of labor and parenting practice in more traditional Asian families. Fathers are held responsible for work and the socialization of older children outside of the home, whereas mothers are responsible for domestic chores and socialization of younger children at home (Costigan & Dorkis, 2006).

Research examining Asian immigrant mothers' child-rearing beliefs and practices in the acculturative context is in its rudimentary stage (Cheah et al., 2013). The effect of acculturation and enculturation on immigrant parents' gendered socialization processes and strategies is not well understood. Research on Asian immigrant women's mothering in single-parent, same-sex or interracial families is almost nonexistent. Much research is needed in this area to explore the potential similarities and differences in mothering experience among and within Asian ethnic groups and to examine varied mothering practices across different family constellations and socioeconomic strata.

Conclusion and clinical implications

As discussed earlier, the pressure to acclimate rapidly to the mainstream culture while finding a job to support their families; the double burden of engaging in full-time employment while staying involved in domestic and parenting responsibilities; and the shifting and renegotiation of gender roles and family hierarchy may render first-generation East Asian immigrant women vulnerable to depression and psychological distress (Dion & Dion, 2001). We posit that feminist therapy or feminist family therapy can be a powerful means of addressing and rectifying the power imbalance created by multiple sources of oppression and hegemony (Sharma, 2001) and that it can represent an effective approach for working with this group. A feminist framework of counseling does not focus on the causes of distress within the individual's own psychic conflicts or problematic thinking and behavior; instead, it considers relevant sociocultural contexts and factors that shape and give rise to the individual's choices and experiences. Although some therapists may fear that adopting a feminist orientation may place their clients at odds with their traditional Asian cultural values and jeopardize their relationships with their families and communities, feminist and multicultural scholars have written about how to effectively help Asian women develop their own voice and self-determination without alienating them from those aspects of culture that are important to their identity (e.g. Kallivayalil, 2007; Kanuha, 2005; Kawahara & Espín, 2007). In the following, we highlight several issues that are specific and pertinent to first-generation East Asian immigrant women's work and family lives.

Asian immigrant women's core work decision – whether to provide financial security for their family through paid employment or to prefer full-time mothering – is often motivated by their desire to enhance their children and family's quality of life (Greenman, 2011). Their sense of self-worth and self-esteem is intricately tied to their family's well-being and their relationships with their social/cultural group (Markus & Kitayama, 1991).

When working with economically disadvantaged Asian immigrant women, therapists should be sensitive of the impact of financial stress, low-prestige jobs, downward mobility and limited social capital on these women's psychological functioning. Feelings of shame, guilt and fear of losing face are common emotions expressed by clients in therapy (Bernstein, Lee, Park & Jyoung, 2007). And for those women who were not employed outside of the home prior to their immigration, it could be particularly stressful for them to take on the role of a "working mother" and "breadwinner" in the US, especially when their husbands are unsupportive or unwilling to share domestic chores and childcare responsibilities. It is important for therapists to become aware of their own biases and judgments about the choices that their clients have to make. Sometimes these choices would mean separating from their children geographically for several years or leaving child-rearing responsibilities to extended family members while working in a foreign land. It is not uncommon for these women to hear reactions such as, "Why bother to have children when you can't be there to raise them!" or "How can you be a good mother when you are so far away from your children?" While it is possible that children who endure prolonged separation from their mothers or parents may experience negative psychological consequences (Ma et al., 2011), therapists should also acknowledge that it is within Asian cultural norms and traditions for children to be jointly raised by grandparents and other extended family members (Bohr & Tse, 2009).

Highly educated and professional Asian immigrant married women tend to have access to social resources such as education, employment, income and decision-making power; however, they may not necessarily enjoy the benefit of an egalitarian relationship. Many of these women still play a primary role in caregiving, homemaking and child-rearing despite their demanding full-time career due to their adherence to traditional gender roles at home. While the fulfillment of role expectations and family responsibilities (including financial contribution) brings satisfaction and enhances one's sense of self-worth and self-esteem (Lee & Mock, 2005), the challenges and difficulties associated with multiple roles may also heighten the risk of psychological problems due to role overload (Noh, Wu, Speechley & Kaspar, 1992).

When exploring how motherhood and paid employment relate to one another, therapists should carefully listen to how these decisions are situated in their clients' narratives of gender-based roles and rules, immigration and acculturative stress, interdependent self-construal and cultural definitions of what it means to be "successful" and to be "a good mother". Therapy should not simply focus on a reduction of emotional distress; it should also help these women give meaning to their choices and affirm their strengths to enact changes in their daily life. In other words, a feminist framework of counseling views women as "active agents of change, rather than victims" (Park, 2008, p. 39) in the process of negotiating their roles and power within a patriarchal family structure as well as within therapy.

Post-migration changes in Asian immigrant women's and men's roles and statuses could additionally burden the women and contribute to tensions and conflicts within the household (Min, 2001). Stress-induced family problems may manifest through their partners' anger displacement; marital stifle; substance use; and acts of domestic violence by their partners, in-laws, and even adult children (Lee, 2007; Rhee, 1997; Sharma, 2001). By explicitly holding and addressing gender and power issues in their minds and actions (Goodrich, 1991), feminist family therapists bring the couple's gendered emotions and behavioral patterns into consciousness. The purpose is to allow each party to examine the unrelenting effect of sociocultural conditioning on their stereotypical interactions (e.g. restricted emotionality in men) and painful feelings and to create opportunities for exploring a new repertoire of emotional experiences and behaviors (Vatcher & Bogo, 2001). More importantly, by being an educator, activist and advocate, therapists can help women break the cycle of conjugal violence by connecting them to culturally and linguistically sensitive career and job retraining programs, providing them with information about their legal rights (e.g. child custody) and partnering with community organizations in order to develop and provide outreach programs for them (Sharma, 2001).

Sociopolitical structures beyond local family and community contexts can perpetuate gender inequality and gender-based oppression. Lack of social policy support and infrastructure, such as unfriendly and hostile work environments, limited flexibility in parents' work schedules and perceived glass-ceiling or wage/salary discrimination may discourage Asian immigrant women from entering and persisting in the labor market. Furthermore, some immigrant women choose to stay at home and endure the challenge of isolated (sole) parenting in order to support their husbands' professional careers and to maintain the family as a safe haven for their children in a racially and ethnically-marginalizing society (Moon, 2003). Therefore, a power analysis in feminist therapy should also consider race and other forms of oppressions that have great impact on Asian immigrant women's family and work lives.

Finally, we would like to point out the dissonance of individualism and collectivism that may occur in the treatment process. Treatment strategies that focus on empowering women should keep in mind Asian cultural values that reinforce traditional gender roles and behaviors. When an Asian immigrant woman chooses to put her family's well-being over her own, asks her young children to step up and help with family responsibilities or perhaps chooses to stay in a power-imbalanced marriage, she should not be viewed as "being resistant to change" or "being an unreasonable parent". Rather, it is important to develop a deep understanding of these women's experiences through a cultural lens, contextualize and validate their choices (Kallivayalil, 2007), de-pathologize their behaviors and interactions – which are often governed by the dominant culture of patriarchy – and bring their diverse voices to the center of therapy as an act of empowerment (Brown, 2010).

References

Addis, M.E. & Cohane, G.H. (2005). Social scientific paradigms of masculinity and their implications for research and practice in men's mental health. *Journal of Clinical Psychology, 61*, pp. 633–647. doi:10.1002/jclp.20099.

Batalova, J. (2011). Asian immigrants in the United States. Retrieved from http://migrationpolicy.org/article/asian-immigrants-united-states.

Bernstein, K.S., Lee, J.-S., Park, S.-Y. & Jyoung, J.-P. (2007). Symptom manifestations and expressions among Korean immigrant women suffering with depression. *Journal of Advanced Nursing, 61*, pp. 393–402.

Blair-Loy, M. (2003). *Competing devotions: Career and family among women executives*. Cambridge, MA: Harvard University Press.

Bohr, Y. & Tse, C. (2009). Satellite babies in transnational families: A study of parents' decision to separate from their infants. *Infant Mental Health Journal, 30*(3), pp. 265–286.

Bronstein, P. (2006). The family environment: Where gender role socialization begins. In J. Worell & C.D. Goodheart (Eds.), *Handbook of girls' and women's psychological health: Gender and well-being across the lifespan* (pp. 262–271). New York: Oxford University Press.

Brown, L. (2010). *Feminist therapy*. Washington, DC: American Psychological Association.

Camarota, S.A. (2012). Immigrants in the United States, 2010: A Profile of America's Foreign-Born Population. Retrieved from http://www.cis.org/2012-profile-of-americas-foreign-born-population#recent.

Chang, T. & Kwan, K.K. (2009). Asian American racial and ethnic identity. In N. Tewari and A.N. Alvarez (Eds.), *Asian American psychology: Current perspectives* (pp. 113–133). New York: Routledge/Taylor & Francis Group.

Chao, R. & Tseng, V. (2002). Parenting of Asians. In M.H. Bornstein (Ed.), *Handbook of parenting: Vol. 4. Social conditions and applied parenting* (2nd ed.) (pp. 59–93). Mahwah, NJ: Lawrence Erlbaum.

Cheah, C.S.L., Leung, C.Y.Y. & Zhou, N. (2013). Understanding "tiger parenting" through the perceptions of Chinese immigrant mothers: Can Chinese and U.S. parenting coexist? *Asian American Journal of Psychology, 4*, pp. 30–40.

Chin, K.-L. (1999). *Smuggled Chinese: Clandestine Immigration to the United States*. Philadelphia: Temple University Press.

Chua, A. (2011). *Battle hymn of the tiger mother*. New York: Penguin.

Chung. R.C. & Bemak, F. (2007). Asian immigrants and refugees. In F.T.L. Leong, A.G. Inman, A. Ebreo, L.H. Yang, L. Kinoshita & M. Fu (Eds.), *Handbook of Asian American psychology* (2nd ed.) (pp. 227–244). Thousand Oaks, CA: Sage.

Cohn, D. (2013). *Demographic Portrait of Adult Children of Immigrants*. Pew Research Center, Washington, D.C. Retrieved from http://www.pewsocialtrends.org/files/2013/02/FINAL_immigrant_generations_re port_2–7-13.pdf.

Collins, P.H. (1994). Shifting the center: race, class and feminist theorizing about motherhood. In E.N. Glenn, G. Chang & L.R. Forcey (Eds.), *Mothering, Ideology and Experience*. New York: Routledge.

Costigan, C.L. & Dokis, D.P. (2006). Relations between parent–child acculturation differences and adjustment within immigrant Chinese families. *Child Development, 77*, pp. 1252–1267. doi:10.1111/j.1467–8624.2006.00932.x.

Dasgupta, S.D. (1998). *A patchwork shawl: Chronicles of South Asian women in America*. New Brunswick, NJ: Rutgers University Press.
Dion, K.K. & Dion, K.L. (2001). Gender and cultural adaptation in immigrant families. *Journal of Social Issues, 57*, pp. 511–521.
Enns, C.Z. (2004). *Feminist theories and feminist psychotherapies: Origins, themes, and diversity* (2nd ed.). New York: Haworth.
Espiritu, Y.L. (1999). Gender and labor in Asian immigrant families. *The American Behavioral Scientist, 42*(4), pp. 628–647.
Fix, M. & Capps, R. (2005). *Immigrant children, urban schools, and the No Child Left Behind Act*. Retrieved from http://www.migrationinformation.org/usfocus/display.cfm?ID=347.
Foner, N. (1998). Benefits and burdens: Immigrant women and work in New York City. *Gender Issues, 16*(4), pp. 5–24.
Foner, N. (2005). *In a new land: A comparative view of immigration*. New York: New York University Press.
Greenman, E. (2011). Asian American-White differences in the effects of motherhood on career outcomes. *Work and Occupations, 38*, pp. 37–67.
Garcia-Coll, C. & Marks, A.K. (2009). *Immigrant stories: Ethnicity and academics in middle childhood*. New York, NY: Oxford University Press.
Goodrich, T.J. (Ed.) (1991). *Women and power: Perspectives for family therapy*. New York: Norton.
Grahame, K.M. (2003). "'For the Family': Asian Immigrant Women's Triple Day." *Journal of Sociology & Social Welfare, 30*(1), pp. 65–90.
Hare-Mustin, R. (1978). A feminist approach to family therapy. *Family Process, 17*, pp. 181–194.
Hays, S. (1996). *The cultural contradictions of motherhood*. New Haven, CT: Yale University Press.
Henderson, A. (2003). Untapped talents: The employment and settlement experiences of Skilled Chinese in New Zealand. In M. Ip (Ed.), *Unfolding history, evolving identity: The Chinese in New Zealand* (pp. 141–164). Auckland: Auckland University Press.
Holland, J.L. (1973). *Making vocational choices: A theory of careers*. Englewood Cliffs, NJ: Prentice Hall.
Hochschild, A.R. (2002). Love and gold. In B. Ehrenreich & A.R. Hochschild (Eds.), *Global woman: Nannies, maids and sex workers in the new economy* (pp. 15–30). New York: Metropolitan Books.
Kanatsu, A. & Chao, R. K. (2008). Asian immigrant fathers as primary caregivers of adolescents. In S. Chuang & R. Moreno (Eds.), *On New Shores: Understanding Immigrant Fathers in North America* (pp. 151–173). Lanham, MD: Lexington Books.
Kanuha, V.K. (2005). Social work practice with women-of-color. In D. Lum (Ed.), *Cultural competence, practice stages, and client systems: A case study approach* (pp. 173–203). Belmont, CA: Brooks/Cole.
Kawahara, D. & Espín, O.M. (Eds.). (2007). *Feminist Reflections in Growth and Transformations: Asian-American Women in Therapy*. New York: Haworth Press.
Kallivayalil, D. (2007). Feminist therapy: Its use and implications for South Asian immigrant survivors of domestic violence. *Women & Therapy, 30*(3–4), pp. 109–127.
Kibria, N. (1993). *Family tightrope*. Princeton, NJ: Princeton University Press.

Kim, B.C. & Ryu, E. (2005) Korean Families. In M. McGoldrick, J. Giordano & N. Garcia-Preto (Eds.), *Ethnicity and family therapy* (3rd ed.) (pp. 349–362). New York: Guilford Press.

Kim, B.S.K., Atkinson, D.R. & Umemoto, D. (2001). Asian cultural values and the Counseling process: Current knowledge and directions for future research. *The Counseling Psychologist, 29,* pp. 570–603.

Kim, K.C. & Kim, S. (1998). Family and work roles of Korean immigrant wives and related experiences. In Y. Song and A. Moon (Eds.), *Korean American women: From tradition to modern* (pp. 103–114). Westport: Praeger.

Kim, S., Conway-Turner, K., Sheriff-Trask, B. & Woolfolk, T. (2006). Reconstructing mothering among Korean immigrant working class women in the United States. *Journal of Comparative Family Studies, 37*(1), pp. 43–58.

Kim, S.Y., Wang, Y., Orozco-Lapray, D., Shen, Y. & Murtuza, M. (2013). Does "tiger parenting" exist? Parenting profiles of Chinese Americans and adolescent developmental outcomes. *Asian American Journal of Psychology, 4,* pp. 7–18.

Kwong, K., Chung, H., Sun, L., Chou, J. & Taylor-Shih, A. (2009). Factors associated with reverse-migration separation among a cohort of low-income Chinese immigrant families in New York City. *Social Work in Health Care, 48*(3), pp. 348–359.

Lee, E. (2007). Domestic violence and risk factors among Korean immigrant women in the United States. *Journal of Family Violence, 22,* pp. 141–149.

Lee, E. & Mock, M.R. (2005). Chinese families. In M. McGoldrick, J. Giordano & N. Garcia-Preto (Eds.), *Ethnicity and family therapy* (3rd ed.) (pp. 302–318). New York: Guilford Press.

Liamputtong, P. (2001). Motherhood and the challenge of immigrant mothers: A personal reflection. *Families in Society, 82,* pp. 195–201.

Loo, C.M. (1998). *Chinese America: Mental health and quality of life in the inner city.* Sage Publication: Thousand Oaks.

Louie, V. (2004). Compelled to excel: Immigration, education, and opportunity among Chinese Americans. Stanford, CA: Stanford University Press.

Lupton, D. (2000). 'A love/hate relationship': The ideals and experiences of first-Time mothers. *Journal of Sociology, 36,* pp. 50–63.

Ma, P-W., Hsu, P., Desai, U. & Zehner, M. (2011, August). The Impact of early separation from parents: Adjustment and parenting concerns among Chinese immigrant families. In C. Yeh (chair) *Parenting and the psychological well-being of Asian immigrant children: The role of acculturation and the immigration context.* Symposium presented at the Annual Convention of the American Psychological Association, Washington, DC.

Markus, H.R. & Kitayama, S. (1991). Culture and the self: Implications for cognition, emotion, and motivation. *Psychological Review, 98,* pp. 224–253.

McCabe, K. (2012). Chinese immigrants in the United States. *Migration Policy Institute.* Retrieved from http://www.migrationinformation.org/usfocus/display.cfm?ID=876.

McGoldrick, M., Anderson, C.M. & Walsh, F. (1989). *Women in families: A framework for family therapy.* New York: Norton.

Moon, S. (2003). Immigration and mothering: case studies from two generation of Korean immigrant women. *Gender and Society, 17,* pp. 840–860.

Noh, S.H., Wu, Z., Speechley, M. & Kaspar, V. (1992). Depression in Korean immigrants in Canada. II. Correlates of gender, work, and marriage. *Journal of Nervous and Mental Disease, 180,* pp. 578–582.

Qin, D.B. (2009). Gender processes of adaptation: Understanding parent–child relations in Chinese immigrant families. *Sex Roles, 60*, pp. 467–481. doi:10.1007/s11199–008–9485–4.

Park, K. (2008). "I can provide for my children": Korean immigrant women's changing perspectives on work the home. *Gender Issues, 25*, pp. 26–42. doi:10.1007/s12147–008–9048–6.

Rhee, S. (1997). Domestic violence in the Korean immigrant family. *Journal of Sociology and Social Welfare*, 24, pp. 63–77.

Rolls, C. & Chamberlain, M. (2004). From east to west: Napalese women's experiences. *International Nursing Review, 51*, pp. 176–184.

Sengupta, S. (1999). Women keep garment jobs by sending babies to China. *New York Times, 1*(4). Retrieved from http://www.nytimes.com/learning/teachers/featured_articles/19990915wednesday.html.

Sharma, A. (2001). Women who have been abused: Improving the effectiveness of Feminist therapeutic interventions with immigrant and racially visible women healing the wounds of domestic abuse. *Violence Against Women, 7*, pp. 1405–1428.

Silverstein, L.B. & Goodrich, T.J. (2003). *Feminist family therapy: Empowerment in social context.* Washington, DC: American Psychological Association.

Smith, A., Lalonde, R.N. & Johnson, S. (2004). Serial migration and its implications for the parent-child relationship: A retrospective analysis of the experiences of children of Caribbean immigrants. *Cultural Diversity and Ethnic Minority Psychology, 10*, pp. 107–122.

Suarez-Orozco, C., Todorova, I.R.A. & Louie, J. (2002). Making up for lost time: The experience of separation and reunification among immigrant families. *Family Process, 41*, pp. 625–643.

Super, D.E. (1957). *The Psychology of Careers*. New York: Harper & Row.

U.S. Census Bureau (2011). 2012 National population projection. Retrieved from http://www.census.gov/population/projections/data/national/2012.html.

U.S. Census Bureau (2012a). Most children younger than 1 are minorities. U.S. Census Bureau News. Retrieved from https://www.census.gov/newsroom/releases/archives/population/cb12–243.html.

U.S. Census Bureau (2012b). Asian/Pacific American Heritage Month: May 2012. U.S. Census Bureau News. Retrieved from https://www.census.gov/newsroom/releases/pdf/cb12ff09_asian.pdf.

U.S. Department of Health and Human Services (2001). *Mental Health: Culture, race, and ethnicity – a supplement to Mental Health: A Report of the Surgeon General.* Rockville, MD: U.S. Department of Health and Human Services.

Vatcher, C. & Bogo, M. (2001). The Feminist/Emotionally focused therapy practice model: An integrated approach for couple therapy. *Journal of Martial and Family Therapy, 27*, pp. 69–83.

Wang, Q. (2006). Relations of maternal style and child self-concept to autobiographical memories in Chinese, Chinese immigrant, and European American 3-year-olds. *Child Development, 77*, pp. 1794–1809. doi:10.1111/j.1467–8624.2006.00974.x.

Wang, Q. (2012). Chinese socialization and emotion talk between mothers and children in native and immigrant Chinese families. *Asian American Journal of Psychology.* Advance online publication. doi:10.1037/a0030868.

Yakushko, O. (2006). Career development of immigrant women. In W.B. Walsh & M.J. Heppner (Eds.), *Handbook of career counseling of women* (2nd ed.) (pp. 387–426). Hillsdale, NJ: Lawrence Erlbaum.

Yip, T., Gee, G.C. & Takeuchi, D.T. (2008). Racial discrimination and psychological distress: The impact of ethnic identity and age among immigrant and United States-born Asian adults. *Developmental Psychology, 44*, pp. 787–800.

Zhou, M. & Nordquist, R. (1994). Work and its place in the lives of immigrant women: Garment workers in New York City's Chinatown. *Applied Behavioral Science Review, 2*(2), pp. 187–211.

4

"Good Asian Moms": Engendering the Model Minority Myth among Indian Immigrant Working Women

Sundari Balan and Ramaswami Mahalingam

Asian Americans have been popularly recognized as one of the most successful groups of immigrants in the United States (Abraham, 2006; Mahalingam, 2006; Pedraza, 2006; Seth, 1995; Zhou, 2002). As a consequence of the Immigration and Nationality Act of 1965, foreign-born Asian migrants have occupied several newly created professional and managerial jobs that also constitute the highest-paying jobs in the American hierarchy (Kanjanapan, 1995; Seth, 1995). Among Asian Indians, who make up the third-largest immigrant group in the United States, more than 77 percent of the foreign-born population is likely to be employed in white-collar jobs; only 58 percent of the white population maintains employment in these types of jobs (Seth, 1995). The relatively poor economic conditions and opportunities for career mobility in their countries of origin have provided further impetus for skilled Asians to migrate to the United States.

The popular belief that Asian immigrants are more hardworking and successful than other minority groups in the United States is commonly known as the "model minority" stereotype (Barringer & Kassebaum, 1990; Hirschman & Wong, 1986). This stereotype pervades many different domains, including gender and familial roles, academic advancement and social mobility. According to the idealized identity model (Mahalingam, 2006), social marginality (i.e. social location vis-à-vis the dominant groups) provides the impetus for the construction, internalization and enactment of the model minority myth in immigrant women's lives in a variety of domains, including motherhood and career. This notion is based on the social construction of differences and how women perceive power differences between their identity groups (e.g. "Asian" mothers) and other identity groups (e.g. "American" mothers) during the post-migration period. Furthermore, the model posits that the endorsement of these idealized identities holds the key to better mental health for women. In addition, the transnational space lends itself to the heightened idealization of such identities by immigrant women themselves. The intersection of these constructs with traditionally gendered spaces, such as motherhood, may be especially ripe for the construction and enactment of idealized identities, especially by immigrants themselves in their own lives (see, for example, Mahalingam, Balan & Molina, 2008).

Power differences are perceived by Asian women not only in relation to women's stereotypes of "American" women and men but also in relation to men in their own families. These are especially intensified when women negotiate multiple intersecting expectations, including being "good" in relation not only to American women and men but also to Asian men and fathers. We expected the confluence of motherhood and work roles to accentuate two types of model minority stereotypes: "good" Asian parents and "successful" Asian workers. A key question that we explored in this study was the following: Which of the two belief systems (i.e. good Asian mother vs. good Asian worker) is related to women's mental health? This question was addressed with the use of an online survey of employed Asian Indian mothers. We also explored how this stereotype may have affected the women's descriptions of their daily lives with their families and particularly with their children. We examined these themes in semi-structured interviews.

Social contexts of immigration, marginality, work and motherhood

Gender scholars have emphasized the social construction of gendered processes, especially for immigrants (Dion, 2006; Espiritu 2001). Both immigration and entry into parenting may intensify these processes. Pessar (1999) noted that the intersecting contexts of gender and immigration produced unique sets of enactments and embodiments for the women residing within them. The same may be said of immigrant Asian working mothers. The traditional intensive mothering ideology, which suggests that mothers should be constantly available and that they are essential to their children's well-being (see, for example, Hays, 1996), is prevalent among both immigrants and non-immigrants alike in the United States. For immigrants, newer pressures provide the impetus for a new type of Asian motherhood. For example, Moon (2003) documented public or shared motherhood, where motherhood extends to women's social networks, particularly grandparents, in many Asian immigrant communities. Latinas who travel to the United States for work engage in transnational motherhood, in which mothers leave their children behind in their home countries and provide for them by working in low-paid occupations abroad (Hondagneu-Sotelo, 1999; Hondagneu-Sotelo & Avila, 1997; Pessar, 1999). These forms of transnational motherhood use existing community resources (e.g. transnational ties) to create newer visions of how motherhood tasks can be accomplished. In doing so, these practices may question dominant ideologies. For example, shared motherhood involves having non-mothers take care of children; this contradicts the dominant intensive mothering ideology, in which mothers are engaged exclusively with their children.

The Asian American model minority stereotype provides additional expectations for women who have young children. The stereotype sets expectations and standards that group members strive to meet and that can also be a source of pressure (Haritatos, 2005; Mahalingam, 2006). For example, Haritatos (2005) noted that the model minority stereotype can be a source of pride for men but that it can be a source of both pride and pressure for women. Thus, for women, the idealized identity appears to have particular salience, and their lives may be particularly complicated by this stereotype. Gender plays a particularly important organizing role in immigrant Asian Indian women's lives. Although families begin their adaptation to the immigration context as a unit, over time, men seemed to turn increasingly toward the work domain to define their identity, whereas women's identity negotiation was shaped by their family's situation and circumstances (Sakamoto, 2006). Adaptation involving their families, their children's schooling, their communities and perhaps even their workplaces begins to organize women's lives. For working mothers, this adaptation is not restricted to being a good mother at home; it also involves being a good employee at work. After all, the model minority stereotype requires the attainment of higher socioeconomic achievement through successful work; in addition, mothers who are conforming to this stereotype must also nurture high-achieving children.

Women often negotiate with these hegemonic, socially constructed identities by consenting to or resisting them through specific belief systems and enactments (Chen, 1999; Glenn, 1991; Hochschild, 1989;

Ong, 1999). Haritatos (2005) noted that Asian Indian and Chinese immigrant women assert their moral superiority to the dominant white group by constructing model minority belief systems (e.g. "Asian American women are more family oriented than white American women are"; "Asian American women outperform white American women in science and math"). The author also observed that these ideals contributed to positive mental health outcomes for the women in her study. Would similar belief systems with respect to motherhood and career be related to mothers' mental health?

Popular culture and media have contributed to the gendering of processes such as motherhood for Asian women. However, few studies have examined the relative role of work expectations. In this study, we focused on Asian Indian immigrant working women who also have children. How are Asian Indian immigrant women's constructions of being a "good" mother and "good" worker related to their endorsement of expectations for the Asian American model minority? How do these beliefs in motherhood and career domains relate to women's mental health? We addressed these questions with the use of a community survey (Study 1).

Study 1

Participants

The study described in this chapter (N = 61) was part of a larger survey of first-generation Asian Indian women who had at least one child less than 16 years old (Balan, 2009). Participants from the larger study who spoke a South Indian language were included for the purpose of this analysis. All participants were first-generation immigrants and were between 18 and 55 years old. All participants were employed for at least 25 hours per week. Participants for the online survey were recruited in a variety of ways (e.g. community agencies and online postings) and were offered $15 for their participation. The average age of the participants was 38 years (standard deviation, 6.63 years), and the mean period of stay in the United States was 12 years.

Measures

Asian American model minority stereotype

The Model Minority Measure developed by Mahalingam and Haritatos (2000) was used in this study; the basic premise of this measure is that Asian Americans are more intelligent than other groups in the United

States. Six items from the measure were used. Respondents were asked to consider their "thoughts about Asian Americans as a group living in the US today" and to rate their agreement with each statement on a response scale from 1 (strongly disagree) to 5 (strongly agree). The alpha was 0.85.

"Good" mother ideal

A modified version of the Asian American Woman Ideal scale (Haritatos, 2005) was used in this study (alpha, 0.85). The original scale assesses the degree to which participants endorse various beliefs about Asian American women in the domains of family and relationships (e.g. "Asian American women have the capacity to make more sacrifices for their family than Asian American men"). For the purposes of the current study, several wording changes were made to the assessment: "women" was changed to "mothers"; "men" was changed to "fathers"; and "Caucasian" was changed to "American". Two items from this measure were used to denote comparisons with "American mothers" (and to capture the stereotypical representation of white women), and two items were used to denote comparisons with Asian American fathers. Respondents were asked to consider their "thoughts about Asian American women and men in general living in the US today" and to rate their agreement with each statement on a response scale from 1 (strongly disagree) to 5 (strongly agree).

"Good" worker ideal

A modified version of the Asian American Woman Ideal scale (Haritatos, 2005) was also used for this part of the study. The scale assessed the degree to which participants endorsed various beliefs about Asian American women in vocational domains (e.g. "Asian American women are better at science and math than white women"). Respondents were asked to consider their "thoughts about Asian American women and men in general living in the US today" and to rate their agreement with each statement on a response scale from 1 (strongly disagree) to 5 (strongly agree). For the purposes of the present study, the scale items were modified so that two items made comparisons between Asian American men and Asian American women, two items made comparisons between Asian American women and American women and two items made comparisons between Asian American women and American men (alpha, 0.74).

Depression

Self-reported depression was measured with the use of the Center for Epidemiologic Studies Depression Scale (Radloff, 1977). This scale measures current levels of depressive symptoms with an emphasis on the affective component of depressed mood (e.g. "I was bothered by things that usually don't bother me"). The alpha was 0.89 in the current study. Participants were asked to think about how often they experienced each symptom during the past week on a scale of 1 (rarely or none of the time [i.e. <1 day]) to 4 (most or all of the time [i.e. 5–7 days]).

Results

Asian model minority and motherhood ideal

We looked at correlations between the Asian American model minority stereotype and two components of the "good" mother ideal: comparisons with Asian American fathers and comparisons with American mothers. We noted that the six items of the Asian American model minority stereotype were positively associated with the relational motherhood ideal comparisons with Asian American fathers ($r = .52$; $p < .001$; $N = 61$). A similar relationship was seen in comparisons with American mothers ($r = .41$; $p < .001$; $N = 61$).

Asian American model minority and worker ideal

We examined correlations between the Asian American model minority stereotype and three components of the career ideal: 1) comparisons with Asian American men; 2) comparisons with American women; and 3) comparisons with American men. We noted that the comparison with American women was significantly related to the Asian American model minority stereotype ($r = .34$; $p < .05$; $N = 61$). Other correlations were not significant.

Ideals and depression

We performed multiple regression analyses to examine the relationships between the various ideals and depression. After controlling for year of birth, the first set of regressions was meant to test which component of the relational motherhood ideal was a significant predictor of depression. We found that comparison with Asian American fathers was a significant predictor of lower depression for the women in our study ($b = -0.34$; $p < .05$). None of the other predictors were significant, and none of the worker ideals were significant.

Discussion

We examined the construction of motherhood and career ideals as a consequence of endorsing the Asian American model minority stereotype. We found that women who endorsed the view that Asian Americans are a model minority also endorsed "good" mother ideals. When we examined worker ideals and the model minority stereotype, we found that the endorsement of the model minority stereotype was positively related to comparisons with American women but not to comparisons with Asian American men or American men. We found that endorsing some components of the "good" mother ideal reduced depressive symptoms. Creating a "we are better" vision probably served to legitimize and reconcile traditional gendered expectations as they contributed to women's own positive mental health. Our study's results are also consistent with previous research findings that showed that the achievement of traditional ideals is important for women's and girls' mental health (Balan & Mahalingam, 2015).

Study 2

First generation Asian immigrant mothers face unique pressures including their internalized moral responsibility for transmitting idealized Asian values to their children. Several studies have suggested that there is considerable heterogeneity with regard to how Asian Indian women respond to these pressures (Inman, Howard & Beaumont, 2007; Kurien, 1999). Londhe (2006) observed that Asian Indian mothers are torn between the child-rearing practices of the home and host contexts. Tummala-Narra (2004) observed that "mothering in a foreign land involves the transformation of one's cultural and maternal identifications". To establish motherhood competence and effectiveness, women also have to separate from their own "maternal figures and their motherlands". During this process, renegotiation occurs, and newer constructions of motherhood are developed.

To understand how women enacted these motherhood ideals in their daily lives, we conducted semi-structured interviews with three women. We examined the narratives of their immigration experiences and of their experiences being mothers with careers in order to assess themes related to the model minority stereotype. An interview protocol was developed to explore various dimensions of motherhood. Each interview lasted for approximately 90 minutes and took place at the convenience of the interviewee. The interviews were tape recorded, and the contents were analyzed for two major sets of themes drawn from the psychological literature regarding mothering and well-being.

The literature that concerns motherhood suggests that a mother's well-being is enhanced when she perceives herself as competent and effective (see, for example, Dumka et al., 1996 and Oakley, 1992). Such a vision of competent parenting is associated with active coping and resilience. We argue that intersections of gender and immigrant cultural identities (e.g. the model minority stereotype) shape the trajectories of becoming a competent mother in unique ways. To understand the unique pressures experienced by immigrant Asian Indian mothers, we began by examining our interview transcripts to better understand the beliefs that women expressed about their own roles as mothers. Any references that these women made to their own roles in their children's lives – including activities that they performed for or with their children – were coded. Because our concern was with model minority beliefs in particular, any references that the women made to expectations regarding their children's achievement and specifically how they perceived their own roles as mothers were coded.

A second dominant theme in the motherhood literature is that of perceived social support (i.e. the role of other people in the motherhood experience). The presence of social support has been considered a source of strength (Oakley, 1992). The literature about Asian Indian mothers (see, for example, Londhe, 2006) suggests that first-generation Asian Indian women have often perceived mothering to be a source of stress, particularly because their social ties are severed during the immigration process. To examine this theme, we explored how women referred to the help that they sought from others for the parenting and upbringing of their children. We coded all references to how women constructed their roles in relation to others. The major themes are enumerated in the following section of this chapter.

Major themes

"I am different from others"

We expected that the theme of achievement would be important for the mothers in our interviews. We explicitly asked our interviewees about their own expectations for themselves as mothers. When talking about her own expectations for her preschooler, Latha, the mother of a three-year-old child, stated the following:

> I don't want her to be a child prodigy or something. Most of the Indians do, right? They push their kids. I am not one of them, I am more easygoing. But I did think of like private versus public schools.

While denying that she focused her attention on her child's achievements, Latha differentiated herself from other members of her group. In doing so, she endorsed the group-specific model minority expectation, yet she resisted it by justifying it when she suggested that she is an "easygoing" person. Psychologically, she resists the idea of the model minority, yet in practice, she does suggest that she may have internalized the achievement expectation of the community by exploring the option of private school for her child, who would only be entering the school system at least two years in the future.

Being different as a source of pride

While Latha denied the achievement emphasis in theory but endorsed it in practice, Meena – a 40-year-old woman with two teenage children – seemed to suggest that she endorsed the belief but that her enactments reflected the reality of her circumstances. When faced with the diagnosis of her younger daughter's learning disability, she contested the dominant view that high academic achievement could be the only source of pride for her as a mother, despite the superior academic performance of her elder daughter:

> When I see her, you know, for me that is a different kind of pride. Because I want to know that, when I die, she is OK and she can take care of herself. Because that really surprises me. Because she has been diagnosed with a lot of little problems, medical problems also.... But she was a very strong kid. So far [as] she is concerned, it is not achievement. Because Nandini, if she reads a book, she gets an A.... With Malati, I don't care whether she gets a B or a C. Of course, I scream at her. But the fact that she can take care of herself is something. Because I am not a tyrant.

While reveling in the academic achievements of her older child, Meena denied that academic achievement was her only source of motherly satisfaction. In doing so, she idealized her younger child's strengths (i.e. "she was a very strong kid"), and she additionally idealized her own role as "not a tyrant". Like Latha, Meena created a vision for herself as a mother: Latha was "easygoing", and Meena was "not a tyrant". In doing so, both mothers resisted the dominant achievement stereotype yet used it as a trope to craft unique and idealized roles for themselves as mothers.

There appeared to be a progression in terms of how the mothers who were chosen for this study presented evidence related to their model minority expectations. Latha seemed to reject the ideology but enacted certain aspects of it. Alternatively, when faced with a unique predicament that forced her to revisit her expectations, Meena seemed to mitigate the impact of the ideology with her practical circumstances. Meena's source of pride seemed to emerge from her resistance to the expectations of idealized motherhood. She attributed her decision to deviate from the achievement expectations to her reading of her child's interests. In response to the choices that they had made, both women idealized their discretionary skills and sensitivity as mothers.

"I am the better parent"

Because the presence and support of others has been suggested to be important for the accomplishment of various motherhood tasks, we examined references to others that our interviewees made with regard to performing tasks in the home. Latha referred to the daily ritual of bathing her daughter:

> We start like...I give her a bath in the evening. Then she will play for some time. Then we try to give her a bath. I give her a bath. Then we read books. One hour. She will say 'Two books' and then it will go to 20. Then we have to read all the books. By the time she falls asleep, it is 9:30, but she is relaxing and in bed. She won't want my husband to read because he will be just...he will read without any expressions. So she will say, 'No, Daddy, I want Mommy to read.' (Laughs.)

Latha also talked about who puts her daughter to bed. She mentioned that she worked farther away from home than her spouse does and hence arrived home much later than her spouse. She began by describing how she and her spouse together give her daughter a bath, but "We" quickly shifted to "I". She further described how she would be the one reading to her child, simply because that is what the child demanded. She attributed it to her partner's inability to respond to the child's needs. Latha took pride in being a better parent by being somehow more effective than her spouse at the performance of daily rituals. During the conversation, she even referred to her daughter's being underfed if she did not return from work on time. The basis for her effectiveness was defined by her reading of her daughter's perceptions in the former case as well as by what was not accomplished when she was absent.

Whereas Latha idealized her own physical presence for the effective accomplishment of parental tasks, Meena, whose teenage daughter struggled socially, tried to help her daughter by creating a network of friends for her:

> Now she says, 'Amma, I don't care if I don't have friends.' All my friends have become her friends. She e-mails them. She'll call them. She'll chat with them. My friends.... So, all of them are her friends. I don't know if it is this society. I don't know if I would have had this problem if I had been in India. Maybe, maybe not.

Meena created a social network that would support not only her but her daughter as well, to the point that her daughter does not miss having her own friends anymore. Meena mentions that she is unsure if she would have had to do this had they lived in India. When referring to the actions that she has taken for her daughter, Meena reflects on her own role: "If she didn't have a mother who took so much initiative for her, she would have gone crazy."

Like Latha, who talked about how her absence is noticed when she is working late, Meena used her imagination to idealize her role as a mother in her daughter's life. Both Meena and Latha created spaces for themselves as mothers in their children's lives, and they viewed themselves as the only ones who could fulfill their children's needs appropriately. Unlike Latha, who was reluctant to call on her spouse because she thought he was ineffective, Meena was able to call upon others, particularly her friends, to support her in her parenting. Latha even expressed reluctance to make use of her network of friends. She had the following to say about calling on her friends to help babysit:

> No, there is no way I am going to go in with.... It's like giving them more work to do.... Everybody has busy lives.... I am sure I will do that if there is an emergency. There are lots of people who can take care of her, even my neighbors can do that.

Meena even valorized her own role as the better parent. When asked what she found challenging about parenting, she had this to say:

> Not me, but my husband. In terms of when she has a tantrum or something... but see, when she has a tantrum or something outside, it is kind of hard. But he loses patience very easily. The more you lose [it] with her, the longer it lasts, but he does not realize that.... Just ignore her or whatever, she calms down very fast.... He also yells at her at that time, that makes it hard. I try to basically distract her,

> away from that mood. He doesn't do that.... Things for her, mostly I will do it. Only if I am busy, he will do.

Meena questioned her spouse's competence with regard to being able to handle any situation that may arise. In doing so, she reiterated the traditional gendered nature of motherhood. A similar choice was made by Lakshmi, another respondent, who had a four-year-old child and an eight-year-old child. Lakshmi had an advanced medical degree, and she chose to work part-time in an emergency ward. She worked on weekends and took late night shifts so that she and her spouse could share all parenting responsibilities. Although her spouse, who held a research position, worked regular hours, Lakshmi chose to work odd hours at locations far away from her home so that she could spend more time with her children.

Role of transnational ties

Latha imagined that having her family available – especially her mother or sister – would have been wonderful during her daughter's early years. However, she mentioned that she had never actually asked her mother or sister to come help her:

> Oh, my God.... It was difficult... because she was in daycare for the first time. She didn't want to go and we were also learning our job and here you don't have the confidence.... It's difficult.... [The] children got chicken pox, and we would take turns... or he would go in the morning and I would go in the afternoon.... But see, those are the times, if you have your mother around or sister around, it helps enormously.

Like Latha, Meena imagined how things might be different had she had access to her parents' help. She mentioned that she would never use the services of a babysitter. Like Lakshmi, who scheduled her work around her children's lives even though it probably caused her great inconvenience, Meena too showed extreme reluctance to use outside help:

> We do everything, once we go home, we do with her. We don't like... like in the weekends or evenings, I never use a babysitter or anything.... I feel guilty doing that.... It will be nice, if we lived in India, where we had our parents close by or something where we could drop them off.... That's how the folks do it here, right?

Although, in theory, Meena felt that it may be nice to have her parents around, she also talked about the conflicts that she faced when her mother visited her. Her mother, who traveled from India to assist her with childcare, originally intended to stay for the maximum allowable time limit for tourist visa holders from India. She decided to leave earlier than scheduled because of the increasing conflicts with her daughter.

> Me and my mom didn't get along very well. They planned to stay for six months. It was that type of a control issue. I guess it is the Indian way. I guess if you are young. She was so upset.

She later remarked that her mother had wondered, "What's the point in me coming here?"

Although our interviewees valued transnational ties over professional help for support, they also demonstrated resistance to use a support system. Lakshmi and Latha attributed the inability of their parents to join them to the practical difficulties of such an arrangement, whereas Meena attributed it to ideological conflicts that she might have with her parents. All three of these mothers created visions of leading different lives in which their own mothers would be able to support them, yet they all showed reluctance to accept their respective mothers' help. At the same time, they prided themselves on being able to cope with their parenting concerns themselves. Latha remarked how she would somehow manage to make macaroni and cheese for her daughter when all of her family members were ill. Meena took time off of work so that she could be with her children during the first year of their lives. Lakshmi scheduled her work so that she could be available to her children. Thus, these mothers all displayed resistance to using family ties even as they continued to endorse such ties at an ideological level.

"I know what it takes to be a good mother"

All of our interviewees expressed confidence in their parenting skills, to the point of being able to give advice to other mothers. Lakshmi stated the following regarding what she thought about other mothers:

> I think every mom should spend time. They do spend time but with their schoolwork. They do not work with them. Those parents think, I am cooking and cleaning, but it is not as effective as spending time with them.

Meena talked about how she would coach her friends on how to teach their kids. She recalled advising one of her friends:

> Don't try to teach her.... Maybe like, whenever you talk about things, 'Eh, look at the red ball.' Don't tell [her], like, 'Come here, this is a red ball; this is blue.' You should not do that. Just your normal, general, whenever you talk to your kids, use colors that way; then they automatically pick [it] up.

Meena had gained confidence in herself as a mother and was now able to advise her friends. In doing so, she idealized her own mothering as superior and as better than that of her counterparts. She claimed that she knew about aspects of parenting that some of her peers did not. Lakshmi asserted that her own priorities with her children (i.e. spending time with them) were more essential to the well-being of the children than accomplishing household chores.

Discussion

In our interviews, we observed that the women idealized and enacted their visions of a "good" mother in a variety of ways. While talking about their own achievement expectations for their children, which is a key component of the model minority stereotype, it is significant to note that our interviewees differentiated themselves from other mothers and even from other "Asian" mothers. By doing so, they resisted the dominant stereotype (i.e. that Asians are overinvested in their parenting roles) while participating in the construction of a valorized role for themselves as super-agentic mothers. As "super mothers", they created a vision that rendered them irreplaceable and essential to their children's positive mental health and well-being. In this way, they negotiated their own self-agency and defined spaces for themselves during the post-immigration phase of their families' adjustment.

The studies discussed in this chapter suggest that newer forms of gendered cultural expectations are related to women's own mental health. In Study 1, we observed that women who believed themselves to be "good" mothers had reduced depressive symptoms. When presented with altered life situations that were inconsistent with the model minority stereotype (e.g. a child with a certain disability who may not achieve the model minority goals of high success and academic achievement), women redefined their social roles and found other ways to "prove" themselves. At times, there appeared to be a chasm between their enactments of motherhood and their beliefs about motherhood.

This was reconciled by attributing the choices made to their own internal attributes (e.g. having the discretionary skills needed to create different expectations for their children or having an "easygoing" attitude). Occasionally, the women valorized the mother's role as compared with the father's role within the post-migration family system. Furthermore, the women valorized themselves and imagined that their absence would be missed. In one case, this belief was legitimized by perceived spousal incompetence with regard to performing routine parental activities.

Women also reconstructed the existing dominant gendered ideology of motherhood (i.e. mothers are more essential than fathers or other family members to children's growth) through culture-specific enactments. With the use of both imagined and real scenarios, they created idealized representations of their own roles as superior mothers who were better than other reference groups. This somehow seemed to contribute to their self-agency and possibly to better mental health.

General discussion

In Study 1, we examined how the endorsement of the model minority stereotype was associated with the endorsement of both the "good" mother and "good" worker ideals. When faced with the pressures of immigration and a new social context that valorized Asian Americans as a "model minority", the women interviewed here seemed to create a model minority vision of motherhood that valorized their own group over others. The embracing of the model minority ideal predicted lower levels of depressive symptoms. The motherhood ideal in particular was associated with lower levels of depression; this suggests that, from a mental health standpoint, motherhood was perhaps more important than career success for these immigrant women.

The quantitative study mentioned in this chapter used a variable-centered approach to understand the relationship between model minority processes, gendered stereotypes (e.g. that of a "good" mother) and depression as a mental health variable. The quantitative section of the study was assigned to examine the significance of empirical relationships among critical variables (see Laub & Sampson, 1998). The qualitative part was conceived to examine the following: 1) the fluidity of person-centered rather than variable-centered practices (see Laub & Sampson, 1998); 2) the degree of ambivalence regarding the endorsement of idealized identities; and 3) the dynamic nature of the construction of motherhood in the women's post-immigration narratives. The selection of participants and the sampling strategy reflected these goals.

Our findings align with the observations in Mahalingam and Rabelo's 2013 study: mixed methods are particularly helpful for studying immigration related psychological processes. However, we must also point to the complexities involved in the study of gendered processes, in which heterogeneity is related to the immediate context (e.g. the mother with the disabled child). When there are intersecting cultural (e.g. immigration) and gendered (e.g. "good" mothers or "good" workers) processes, women's mental health is shaped by larger power expectations; this was reflected by depressive symptoms in the studies discussed here. We found that the "good" mother ideal involved lowered depression for the women in our study but that the "good" worker ideal did not, although both ideals were related to the general model minority stereotype. By taking a person-centric approach, we attempted to stop creating "caricatures of immigrants" (see Mahalingam & Rabelo, 2013). We found that women were self-aware of ethnic and gender expectations, such as those associated with the model minority stereotype. They also resisted such expectations and reconstructed new visions and meanings of parenthood, which appeared to demonstrate self-agency and an active coping strategy.

Limitations and future directions

The studies discussed in this chapter have several limitations. What is presented here is a limited analysis of the belief system of a small number of survey respondents. Our convenience sampling approach may have obliterated regional and context specific variations. Furthermore, religion and caste variations should also be considered in future studies. More studies are needed to understand how these expectations are related to the model minority stereotype in other Asian subgroups as well as longitudinally related to it, and it is also possible that these results would change as children grow. We could not account for the variations in the professional engagement of the women studied. Perhaps the women who perceived their careers to be successful demonstrated differential associations with depressive symptoms.

Conclusions

We examined the relationship between the endorsement of the Asian American model minority belief system and the "good" mother and "good" worker ideals. Our survey results provided support for this relationship. In addition, only the "good" mother ideal was related to lower depression. In our qualitative study, we observed that women both accepted and resisted this ideal, which sometimes created dichotomies between their own beliefs and enactments. We found considerable heterogeneity with regard to how women enacted the model minority stereotype; they defined themselves as "good" and at times viewed themselves as indispensable "super parents" in a wide variety of domains (e.g. as compared with their spouses or with other Asian parents). This belief system appeared to be a source of motherhood self-agency and active coping in the face of intersecting gender and immigrant cultural expectations.

References

Abraham, M. (2006). Model minority and marital violence: South Asian immigrants in the United States. In R. Mahalingam (Ed.), *Cultural psychology of immigrants*. Mahwah, NJ: Lawrence Erlbaum.

Balan, S. (2009). Being Asians, good moms and great workers: Investigating the psychological contours of Asian Indian women's model minority experience. Doctoral Dissertation, University of Michigan, Ann Arbor. Retrieved from http://deepblue.lib.umich.edu/bitstream/handle/2027.42/62404/sundarI_1.pdf?sequence=1.

Balan, S. & Mahalingam, R. (2015) Family context of well-being among married women and men within extreme son preference and patriarchy. (Manuscript submitted for publication).

Barringer, H. & Kassebaum, G. (1989). Asian Indians as a minority in the United States: The effect of education, occupations and gender on income. *Sociological Perspectives, 32*, pp. 501–520.

Chen, A.S. (1999). Lives at the center of the periphery, lives at the periphery of the center: Chinese American masculinities and bargaining with hegemony. *Gender & Society, 13*, pp. 584–607.

Dion, K. (2006). On the development of identity: Perspectives from immigrant families. In R. Mahalingam (Ed.), *Cultural psychology of immigrants*. Mahwah, NJ: Lawrence Erlbaum.

Dumka, L.E., Stoerzinger, H.D., Jackson, K.M. & Roosa, M.W. (1996). Examination of the cross-cultural and cross-language equivalence of the parenting self-agency measure. *Family Relations, 45*, pp. 216–222.

Espiritu, Y.L. (2001). "We don't sleep around like White girls do": Family, culture and gender in Filipina American lives. *Signs, 26*, pp. 415–440.

Glenn, E.N. (1983). Split household, small producer and dual wage earner: An Analysis of Chinese-American family strategies. *Journal of Marriage and the Family, 45*, pp. 35–46.

Glenn, E.N. (1991). Cleaning up/kept down: A historical perspective on racial inequality in women's work. *Stanford Law Review, 43*, pp. 1333–1356.

Haritatos, J. (2005). *"Becoming American:" The intersections of cultural, ethnic, and gender ideals in predicting levels of perceived stress and mental and physical health among Asian Immigrants*. Unpublished Doctoral Dissertation. University of Michigan.

Hays, S. (1996). *The cultural contradictions of motherhood*. New Haven, CT: Yale University.

Hirschman, C. & Wong, M.G. (1986). The extraordinary educational attainment of Asian-Americans: A search for historical evidence and explanations. *Social Forces, 65,* pp. 1–27.

Hochschild, A. (1989). *The second shift.* New York: Viking.

Hondagneu-Sotelo, P. (1999). Introduction: Gender and contemporary U.S. immigration. *The American Behavioral Scientist, 42,* pp. 565–576.

Hondagneu-Sotelo, P. & Avila, E. (1997). I'm here but I'm there: The meanings of Latina Transnational Motherhood. *Gender and Society, 11,* pp. 548–571.

Inman, A.G (2006). South Asian women: Identities and conflicts. *Cultural Diversity and Ethnic Minority Psychology, 12,* pp. 306–319.

Inman, A.G., Howard, E.E. & Beaumont, R.L. (2007). Cultural transmission: Influence of contextual factors in Asian Indian immigrant parents' experiences. *Journal of Counseling Psychology, 54,* pp. 93–100.

Kanjanapan, W. (1995). The immigration of Asian professionals to the United States: 1988–1990. *International Migration Review, 29,* pp. 7–32.

Kurien, P. (1999). Gendered ethnicity: Creating a Hindu Indian identity in the United States. *The American Behavioral Scientist, 42,* pp. 648–670.

Laub, J. H. & Sampson, R. J. (1998). Integrating quantitative and qualitative data. In G. H. Elder, Jr. & J. Ziele (Ed.), *Methods of life course research: Qualitative and quantitative approaches* (pp. 213–230). Thousand Oaks, CA: Sage.

Londhe, R. (2006). Decisions about child-rearing practices. A study of Asian Indians in the United States. Unpublished Doctoral Dissertation. University of Connecticut, USA.

Mahalingam, R. & Haritatos, J. (2000). Development of a scale measure of model minority for Asian Americans. Unpublished manuscript. University of Michigan.

Mahalingam, R. & Leu, J. (2005) Culture, essentialism, immigration and representations of gender. *Theory & Psychology, 15*(6), pp. 841–862.

Mahalingam, R. (2006). Cultural psychology of immigrants: An introduction. In R. Mahalingam (Ed.), *Cultural psychology of immigrants.* Mahwah, NJ: Lawrence Erlbaum.

Mahalingam, R., Balan, S. & Molina, K. (2008). Transnational motherhood and families: An intersectionality perspective. In S.A. Lloyd, A.L. Few & K.R. Allen (Eds.), *Handbook of feminist family studies.* Thousand Oaks, CA: Sage.

Mahalingam, R. & Rabelo, V.C. (2013). Theoretical, methodological and ethical challenges to the study of immigrants: Perils and possibilities. In M.G. Hernández, J. Nguyen, C.L. Saetermoe & C. Suárez-Orozco (Eds.), *New directions in child development: Frameworks and Ethics for Research with immigrants.* San Francisco, CA: Jossey-Bass.

Moon, S. (2003). Immigration and mothering: Case studies from two generations of Korean immigrant women. *Gender and Society, 17,* pp. 840–860.

Mullings, L. & Wali, A. (2000). *Stress and resilience: The social context of reproduction in central Harlem.* New York: Kluwer Academic Publishers.

Oakley, A. (1992). Social support and motherhood: The natural history of a research project. Oxford, UK: Blackwell.

Ong, A. (1999). Cultural citizenship as subject making: Immigrants negotiate racial and cultural boundaries in the United States. In R.D. Torres, L.F. Miron & J.X. Inda (Eds.), *Race, ethnicity and citizenship: A reader* (pp. 262–294). Oxford: Blackwell.

Park, K. (2007). Constructing transnational identities without leaving home: Korean immigrant women's cognitive border-crossing. *Sociological Forum, 22,* pp. 200–218.

Pedraza, S. (2006). Assimilation or transnationalism? Conceptual models of the immigrant experience in America. In R. Mahalingam (Ed.), *Cultural psychology of immigrants*. Mahwah, NJ: Lawrence Erlbaum.

Pesar, P.R. (1999). Engendering migration studies: The case of new immigrants in the United States. *American Behavioral Scientist, 42,* pp. 577–600.

Radloff, L.S. (1977). The CES-D scale: A self-report depression scale for research in the general population. *Applied Psychological Measurement, 1,* pp. 385–401.

Sakamoto, I. (2006). A model of cultural negotiation and the family: Experiences of Japanese academic migrants in the US. In R. Mahalingam (Ed.), *Cultural psychology of immigrants*. Mahwah, NJ.: Lawrence Erlbaum.

Seth, M. (1995). Asian Indian Americans. In P.G. Min (Ed.). *Asian Americans: Contemporary trends and issues*. Thousand Oaks, USA: Sage.

Tummala-Narra, P. (2004). Mothering in a foreign land. *The American Journal of Psychoanalysis, 64,* pp. 167–182.

Wong, M.G. & Hirschman, C. (1983). Labor force participation and socio-economic attainment of Asian-American women. *Sociological Perspectives, 26,* pp. 423–446.

Wong, M.G. (1986). Post-1965 Asian immigrants: Where do they come from, where are they now, and where are they going? *Annals of the American Academy of the Political and Social Science, 487,* pp. 150–168.

Zhou, M. (2002). Contemporary female immigration to the United States: A demographic profile. In P. Strum. & D. Tarantolo (Eds.), *Women immigrants in the United States* (pp. 23–34). Washington DC: Woodrow Wilson International Center for Scholars.

5

Being a Woman, a Muslim, and an Afghan in the USA: Dilemmas of Displacements

Huma Ahmed-Ghosh

In this chapter I discuss two interviews with Afghan mothers who live in Southern California. The purpose of the interviews was to understand their gender and ethnic identity issues as they raise their children in an environment that is hostile to Muslims and Afghans. What makes the position of Afghans as refugees unique in the USA is that they are fleeing their home country of Afghanistan to become refugees in the very country that is at so-called war with theirs. This is further complicated by post-9/11 hostilities, stereotypes and suspicion of Muslims in general and Afghans in particular, leading to their marginalization in USA. For refugees in general, one cannot discuss their issues, adjustment or lack thereof to the USA as a homogenous group. Their experiences, lifestyles and expectations from their families, their communities and the larger society are determined by their gender, class, ethnicity and age as well as, especially for Afghans, their wave of migration and the US-Afghan conflict.

Although refugees and all displaced persons may encounter similar obstacles in flight and during displacement, they are not a homogenous group. They carry with them different experiences, histories and reasons for flight. In addition to these differences, male and female refugees experience displacement in significantly different ways. Displacement is a difficult journey for women. As presumed caregivers, women travel with children or other family members in need of assistance, causing them to prioritize others' needs ahead of their own. In this chapter, using ethnographic analysis, I discuss how geopolitical forces impact Afghan mothering in San Diego.

Migration patterns are based on "push" and "pull" factors. Basically, push factors refer to reasons why people leave a country, such as political persecution, economics, social discrimination, environmental crisis and war. Pull factors refer to reasons that attract people to migrate because of better economic and social opportunities. Pull factors are commonly associated with an economically developed country encouraging skilled labor and unskilled labor to enter their country to boost their economy; an example would be the US granting entry visas to technologically skilled immigrants from Asia and Europe. In recent decades, conflicts have provided the largest reason for push factors.

Here I will deconstruct the lives of two mothers who see themselves as products of migration patterns that are diverse. One mother sees herself as an immigrant having arrived in the US in the late 1970s due to a mix of push and pull factors, whereas the other mother sees herself as a refugee entirely due to push factors. Post-9/11 both women wrestle with the issues of "home" and of retaining their faith of Islam as part of preserving their Afghan culture.

The next section briefly lays out some definitions of refugees and the impact of forced migration on gender hierarchies.

Refugees and migration

Populations who flee their country of origin are referred to as refugees. In 1951, the Convention Relating to the Status of Refugees defined who is a refugee. A refugee is a person who

> owing to a well-founded fear of being persecuted for reasons of race, religion, nationality, membership of a particular social group or political opinion, is outside the country of his [her] nationality and is unable or, owing to such fear, is unwilling to avail himself [herself] of the protection of that country. (UNHCR, 1951 Convention, 1967 Protocol)

Similarly, the US Immigration and Nationality Act of 1965 defines a refugee as a person who "is unable or unwilling to return to his or her country of nationality because of persecution or a well-founded fear of persecution on account of race, religion, nationality, membership in a particular social group, or political opinion" (Martin and Yankay, 2014, p. 1). This definition qualifies Afghan immigrants to seek refugee status in the USA.

Refugee migration is, therefore, forced migration, and as Matsuoka and Sorenson (1999) point out, forced migration can be traumatic because of the new environment that refugees have to adapt to and the negotiations and cultural confusion that often lead to "assuming new roles and expectations, behaviors, and relationships" different from those "that have operated in the past." McSpadden (1999) adds another layer to this process by discussing how the "resettlement experience is powerfully shaped by governmental policies and programs and informed by the cultural values and norms of the receiving society" (p. 244). These adaptations have severe repercussions on gender relationships, many times adverse, though sometimes they do have positive outcomes.

For all immigrants and refugees, the process of migration and cultural adaptation to the new environments is highly gendered (e.g. Espín, 1999; Kallivayalil, 2004). I draw upon Doreen Indra's (1999) work to focus on the specific cultural context of gender relationships among Afghan refugees. Indra describes gender relationships as relational and contextual. She connects relationships and shifting hierarchies not just among individual men and women but also in the institutions and cultures that keep these relationships fluid. Refugee women's lives are thus impacted not just by forced migration but also by gender hierarchies they knew before and those they now encounter in the new country as well as by cultural institutions in the new place. The impact of all these processes is also affected by their socioeconomic and political standing in the country they immigrate to. These variables are very pertinent to Afghan refugees in the USA because of the reception they receive in the US and the perceptions of Afghan Muslim women that frequently need to be countered.

Issues of identity, then, for Afghan women who are mothers in the US diaspora are complex and nuanced because they navigate individual and collective identities in their attempts to cope with a foreign culture. All immigrant and refugee families have to adjust to new patterns while trying to preserve what they consider essential in their backgrounds (e.g. Foner, 1997). In many instances, in their efforts to replace the loss that is experienced through migration, community identity may take precedence over individual identity, as we will see in the interviews reported in this chapter. As Mortland (1994) concludes, a strong ethnic identity proves powerful to contest stereotyping by the majority and could be interpreted as a political act.

In the following section, I will briefly outline the methodology I used in this research and then briefly lay out the history of the three waves of migration from Afghanistan in recent decades in order to situate Afghan mothers' experiences and then embark on the analysis of my interviews to highlight Afghan mothers' dilemmas as they raise their children in the USA, which is still at "war" with their home country.

Method

Since 2003, I have visited Afghanistan half a dozen times. My research agenda looks at how Afghan-run women's nongovernmental organizations (NGOs) strategize women's rights within an Islamic state. This research interest led me to seek out the Afghan community in San Diego. Through networking with a local organization that worked in Afghanistan and with some Afghans on panel discussions about gender and war in Afghanistan that I participated in, I made contacts with the community. In Kabul, I met families of some of the refugees in San Diego who further strengthened my connections with the community. Through casual conversations at Afghan parties and through my efforts to find jobs for Afghan women, I learned some important details. For example, I realized that on many occasions Afghan women were hesitant to go out of the house because of what they perceived as Islamophobia in the larger population and a sense of distrust toward them. These conversations further led to discussions about the dilemmas of how to inculcate Afghan values in their children in a hostile environment. Such concerns led to the idea of this research.

My sample consisted of twelve Afghan women who were mothers. Four mothers who came to the USA before the 9/11 event in the first wave of migration (discussed later) and eight women who came to the USA after 9/11 in the third wave of migration. The ages of the women who came in the first wave ranged from 55 to 65. The women interviewed from the third wave ranges from 30 to 40. All the women had achieved high school education with half of them having completed their BAs in Afghanistan or in Pakistan as refugees. The participants also came from middle-class to upper-middle-class backgrounds, making it possible for them to migrate to the USA. The sample was collected through snow-balling once I discussed the project with two Afghan women whom I came to know quite well over the years. I interviewed each woman for at least 2–3 hours twice in her own home. The mornings were most convenient to interview these women because husbands had left for work and children were in school making it easier for the women to talk openly about her experiences. The interviews were conversational, with open-ended questionnaires and detailed note taking. I did not use a tape recorder, because the women expressed reservations about being on tape and did not want some of the information they were giving me to be published. I was able to base my research on participation observation also because I attended their family and community events. Broadly, my questions pertained to their journey and experiences of migration to the USA, specifically as women; their cultural adjustment process and relationships with the larger community; and the extent to which Islam played a role in their lives as immigrants and refugees. Focused questions related to their roles as mothers in Afghanistan and in the USA, differences in mothering as compared to their mothers' styles, differences in bringing children of different ages to the USA and between children born before arriving in the US and those born here. I also engaged some of the men in casual conversations about their role in their families to see if they perceived any cultural changes in men as husbands and fathers.

This chapter is based on a small subsample of the women I interviewed. For this chapter, I selected just two women in order to highlight the differences in mothering among Afghans. I base my discussion on their expressed opinions and their time of arrival to the USA. Both women reflect the general sentiments, identity issues and mothering concerns of their migratory and age groups. I have given pseudonyms to the participants in this research. The two Afghan mothers I write about in this chapter are Shahnaz and Seema. Nadira, the adult daughter of Shahnaz, facilitated the translation and interpretation of her mother's interview as well as provided some background information on their migration issues. The attempt here is not to generalize about Afghan mothers' experiences in diaspora through a homogenizing analysis but to tell the story of two women who raise pertinent issues of cultural adjustments in a politicized diaspora and a US context which is currently hyper-sensitive to Muslims and Afghans.

Afghan migration waves

Immigration by waves and social class plays a very important role in an Afghan mother's location in the USA.

The first wave of emigration for Afghans occurred when the Soviets occupied Afghanistan in 1979. Those Afghans who had invested or banked their money outside their country had enough resources to buy an air ticket and leave for the West; middle-class Afghans moved to India, and those with fewer resources walked to Iran and Pakistan. The second wave of migration occurred in 1992 when the Mujahideen took over Afghanistan with assistance from the USA.

Mujahideen means freedom fighters. They are Afghans who fought the Russians and were supported by the USA, Pakistan and Saudi Arabia through arms and other financial assistance.

The third wave of Afghan refugees occurred in 1996 when the Taliban took control of the government. The name Taliban refers to students of religion. The Taliban were mainly Afghans who grew up in Pakistani refugee camps and came into power in Afghanistan to quell the civil war during the Mujahideen regime. They were also initially supported by the USA.

The fourth wave of migration from Afghanistan started on October 7, 2001 when the USA started bombing the country. Post-9/11, emigration of men from Afghanistan has been limited mainly due to denial of visas to young men. Since then, internal migrations and remigrations from one refugee camp to another still continues for Afghans. Such is the case with the post-9/11 refugees who have been shuffled from country to country before relocating to the USA.

In this chapter, I will talk about Afghan women, specifically mothers, because their gendered experience is heightened by the very process of migration and remigration or more precisely by repeated dislocations. Hence, coming to the US as adults, Afghan women face a harsh reality of alienation, suspicion and a very different culture. They do not necessarily speak English or understand Western culture.

Social disruption caused by wars has created contradictory spaces for refugee Afghan women in the USA where many have had to look for work (coming from backgrounds where they did not leave the house) while simultaneously limiting their social spaces with controls over their dress and lifestyle. Even as greater cultural control over women's bodies and sexuality is exercised by the community in the name of national honor, limited economic resources compel men to accept that women need to have some access to the public space. Experiences of motherhood are also constantly shifting, adjusting and morphing given their dislocations and new sociocultural and socioeconomic environment.

In the following sections, I will detail the interviewees' responses to topics on kinship and cultural adaptation as migrants and refugees. I will later analyze their responses through the concept of "home".

Afghan fathers

Before I proceed, I will engage in a brief discussion on shifting Afghan fathers' identities in the USA. But I do want to reiterate that while there may be a visible shift in gendered performances in the household due to migration, at no point is the gendered hierarchy based on indigenous patriarchal norms disrupted. Men continue to be seen as the heads of households, treated reverentially, privileged in the household through caregiving by both wife and children and prioritized through serving of food and other comforts.

One cannot talk about Afghan mothering without mentioning fatherhood, especially among refugees and immigrants where these roles are traditionally defined by indigenous cultures and extended households.

The discussion on fatherhood is vital because it is reflected in the mothering process for migrants and refugees. The breakdown of both gendered cultural expectations and indigenous kinship structures redefines gender roles within their new country, leading to differential burdens for both men and women.

As refugee men, especially from son-centered countries where they are entitled to privileges from birth just for being born male, the loss of status and stature in this new country leads to changing behavioral patterns and interpersonal relationships within the newly formed family structure. Afghan men start experiencing racism and Islamophobia since many are recognizable through beards as Muslims and some through their brown skin and accents. This is problematic for many men because besides coming from gender-entitled backgrounds in Afghanistan, they also come from economically and socially privileged backgrounds, but they have lost that social and economic standing through their migration journeys. Added to this disempowerment is a sense of emasculation because of their inability to sufficiently provide economically for their families in the USA. Many Afghans have relocated from previous refugee camps in Pakistan, India and Iran where they were unable to acquire education and employment skills. Language barriers make it harder for them to find "decent" jobs in the USA.

This is further compounded by their lack of participation in the protection of their homeland from foreign invasions. Migrating to the USA has also added to their guilt and inability to fight for their country against the occupiers. While living in the USA, Afghan men have to confront the masculinist discourse of the West which emasculates the Afghan man by constantly showing in the media the deteriorating status of women in Afghanistan. Repeated images of "oppressed" veiled women, women in poverty, girls not attending school, and honor killings of Afghan women reinforce the western perception of a culture that is dictated by a patriarchy that is entrenched in "backwardness" in contrast to a western, liberal and gender-equality based one. This projection is then extended to the inability of Afghan men to respect and take care of women, leading to the ultimate need by the USA to go to war to "liberate women".

Simultaneously, the Afghan father has to assume parenting roles that he would not have engaged in necessarily in Afghanistan. For example, due to women's lack of knowing English, lack of mobility because they do not drive and/or are veiled and lack of self-confidence in public communication, Afghan fathers have had to negotiate the so-called private/public sphere on behalf of their wives and children. Examples are fathers' participation in the children's schools through admissions, homework, attendance of parent-teacher meetings and negotiations with children's doctors. If their children engage in extracurricular activities, these also become the father's responsibility. Fathers are called upon to purchase groceries, do yard work and perhaps perform other household tasks. These new realities heighten the women's dependencies on their husbands while also physically and psychologically dealing with their husband's loss of status in the larger Western society. The rearrangement of family dynamics negatively impacts both mothers and fathers and places a bigger burden on the mothering process.

As I will discuss later in this chapter, the role of children within these diasporic households also shifts compared to their roles before immigration to the US. As mentioned earlier, gender hierarchies within the home are still firmly in place, creating a bigger burden on women who have to preserve the gendered cultural expectations with limited resources, without household help or kin group for support and with limited mobility and acculturation tools. In other words, migration can cause a flux in gendered hierarchies that can be problematic, although at times this can be positive.

Kinship and mothering

One of the major setbacks for Afghan refugees, specifically women as mothers, is the breakdown in the kinship system. I should point out that there are as many disadvantages for women who live in extended households as there are advantages, but from the interviews it was apparent that in a foreign land the lack of those familial and social networking support systems did place burdens on mothers. The interviews showed that this breakdown was reflected in different ways.

Both participants in this research, Shahnaz and Seema, waxed eloquent about "back home" as a "golden era" that provided "conveniences and comforts". There may be some truth to this, but for many migrants, especially refugees, such talk is indulged in as a coping strategy in an attempt to maintain one's culture and group identity in an alien environment.

Negotiating migration: Shahnaz and Nadira

Nadira, a 37-year-old Afghan immigrant, interpreted her mother Shahnaz's, interview and elaborated on many issues garnered from the interview. Nadira mentioned that when they migrated to the USA from Kabul in 1980 – the first wave of migration – they deliberately landed in Los Angeles because her father had some relatives there. Nadira and her mother prefer to refer to themselves as migrants rather than refugees. They prefer the former because they left "willingly" since they had an opportunity to migrate, even though the Russian occupation had occurred and many people were fleeing Kabul. Shahnaz's brother-in-law had migrated to the USA a decade before the Russian occupation and was able to find her husband a job in Los Angeles. Shahnaz's family lived with their relatives in Los Angeles on arrival to the USA as an extended household. While there was a bit of decline in their class status compared to their wealth in Kabul, Shahnaz's family continued to have a relatively comfortable life in the USA. Coming from a secular-left political background, the family was not very religious and therefore adopted Western ways easily. The upper-middle-class and elite in Kabul were quite westernized in dress, education, and so on. Therefore, the shift to the USA was not such a big culture shock. Those who were in the first wave of migration in the early 1980s and had the means to move to the West had less traumatized lives and cultural adjustment issues because they already came from modernized families in Afghanistan.

Complicated kin advantages and disadvantages

Despite their easier adaptation, the sense of home remained a major issue for Shahnaz. For one, she had an independent household in Kabul and did not have to live with her in-laws. This, she felt, gave her freedom to run her household and bring up her children on her own terms without "interference" from the in-laws. Although they lived in a single family home in Kabul, the Shahnaz's relatives, sisters and parents lived nearby and helped out with the children and other chores, creating a pseudo-extended household. Shahnaz also had a couple of servants at her home who did the bulk of the housework. Mothering, she said, was not seen by her as a sole responsibility for her children. Neither did she consider it burdensome, nor was she dependent on her husband to help her out with household chores and childcare.

On their move to the USA, Shahnaz had to adjust to a very different lifestyle because for the first time she physically lived in an extended household and most importantly had to engage in household chores because of the lack of servants. Shahnaz was not fond of those relatives, because they treated her differently since she was from a different ethnic community, but mothering the young children was easier, in retrospect, because they were part of an extended household and an Afghan community. The domestic chores were divided between the adult women. The family ate Afghan food every day, and they all spoke Dari and visited the mosque once in a while. They mainly socialized with other Afghan families; this helped Shahnaz retain her Afghan social status and eased her process of parenting.

Cultural alienation and a sense of loss

When the situation in the extended household became too tense and Shahnaz started feeling uncomfortable, the family moved to San Diego. Here in the cultural and social anonymity of a neighborhood that was basically a white middle-class neighborhood, Shahnaz was totally lost. After initial culture shocks, the children adjusted to the neighborhood and school. They did not have many Afghan friends, and they socialized mainly with the diverse school friends. Their economic class also suffered because both parents had to work hard to provide for the family. According to Nadira, Shahnaz's daughter, this led her two younger brothers to "stray away" from their roots and identify more with the African-American children in school, leading to a lifestyle that the parents could not comprehend.

Shahnaz was disappointed that they could not afford a house in a "better" neighborhood because she was prejudiced against blacks and was embarrassed to be in a working-class neighborhood. Shahnaz felt she had failed her children: she felt that their "straying off" was because she was not home enough to take care of them. She felt the lack of family support and the lack of extended kin members. Even though they chose to move out of the extended household in Los Angeles, they still felt the pressure of being judged by their peers.

In San Diego, the Afghan community was very small and scattered in the early 1980s and different from Shahnaz's family who was liberal and did not practice Islam or require their women to veil. When Shahnaz felt she could not cope, she wanted to move back to Los Angeles, but the children did not. Visits to Los Angeles sometimes for an Afghan holiday or other occasions became less frequent when the children felt that a cultural and educational gap was growing between them and their cousins.

In San Diego, Shahnaz has tried to negotiate with her children whose new culture she found alienating. Her frequent attempts to instill Afghan cultural values of respect for the elders, obedience to parents and single-minded devotion to academics were regularly met with resistance. Such discussions and coercions would end up with the mother and father expressing their frustrations through soliloquies on how wonderful their upbringing had been, how great Afghanistan was and how the extended household was equally responsible in the enculturation of children. Shahnaz's husband, though very appreciative of all the work she did, was not able to help much with household work, due to his heart ailment. He died seven years ago, continuing to yearn to go back to Kabul.

This led to Shahnaz's depending more and more on her older daughter to communicate with her sons. Nadira feels that there is a gender divide in the socialization of children. As the children started growing up, this gender divide became quite apparent. Clothing became an issue for the daughters. Daughters were expected to preserve Afghan culture and their Muslim identity through relatively moderate dressing.

Interestingly, Shahnaz has become more traditional in her mothering practices. When I questioned her on it, she claimed that now that "the Americans" are stereotyping them through their perceptions on Islam it was her duty to preserve her religious identity, even though she did not strictly adhere to the faith. Shahnaz insisted that her daughters dress "decently", not date and not stay out late at parties. These restrictions were not imposed on the sons. Post-9/11, Shahnaz saw a strong need to identify with liberal Islam, and post-2001, she felt closer affinity with Afghan refugees – though she was still hesitant to identify with them as a refugee. Shahnaz gradually started drawing Nadira into her mothering realm. She was now asked to co-mother her younger siblings because of her ability to negotiate with the outside "American" world, the school system, and so on. This co-mothering was earlier provided by other female members of the extended household in Kabul, but in the USA family responsibilities have been rearranged. The burden of housework and care of younger children now falls on Nadira's shoulders.

Shahnaz kept reminiscing about her days in Kabul, about how they had more women in the household as well as servants. She kept saying how lucky her mother was in Kabul because she had such few responsibilities in running the household and rearing the children. Of course, their class background in Kabul also played an important role in mothering activities and identities. The kin group was frequently evoked to talk about strong family ties and sharing of familial responsibilities; kin groups were also evoked to talk about the sense of duties and sacrifices family members made to uphold the extended household and their honor.

Nadira's mother made her first trip to Kabul last year after 30 years of absence and came back with a sense of resignation that her home now is San Diego. The war had changed everything in Afghanistan. Kabul was unrecognizable, and her family and friends had left. Now Shahnaz resorts to Afghan and Turkish TV watching, Afghan food and an insistence on Dari being spoken at home – ensuring that her children, all adults now, do not lose the one strong Afghan bond she wants to maintain with them.

Seema: Conflicted and forsaken

Seema came to the USA from Pakistan, having migrated there in 1992. She and her husband's family had left Afghanistan when the Mujahideen took over Afghanistan, and the ensuing civil war led to a second wave of migration. Seema's family moved to Pakistan and lived in a refugee camp for some months. Later, they found a very tiny house in which eight members of her family lived.

She married her husband in Pakistan where he worked hard to make ends meet while also going to college and finally attained an undergraduate degree and a certificate in business. But similar to most immigrants and refugees from developing countries, his degrees were not recognized in the USA, forcing him to have multiple low-paying jobs. Currently, Seema's husband is pursuing higher education while working full time. In Pakistan they had two daughters. When the opportunity arose, their daughters migrated to the USA as refugees.

Seema came from an upper-middle-class family that lost much of their wealth when they moved to Pakistan. In Afghanistan, they had to leave their house, land and possessions and flee their town overnight. The family was able to marry three sisters to men from similar class backgrounds. The younger sons, though, grew up mainly in the refugee camps and did not receive adequate education, and they therefore could not find decent jobs.

Seema's in-laws who had also temporarily moved to Pakistan decided to go back to Kabul, despite the harsh conditions, because they just could not adapt to the refugee camps and "preferred death" to the humiliating living conditions. While her father-in-law returned to his ancestral property, he had no job. His refugee sons living in the Netherlands and the USA had to remit money for living expenses to family members in Kabul, further adding to Seema's economically strapped situation.

Seema's experiences of migration differ greatly from Shahnaz's because Seema's family moved from one refugee camp to another. On this journey, Seema's family lost its economic resources: her husband was not able to achieve the level of education he would have wanted and came to the USA as refugees staying in low income housing. This was a culture shock to Seema because now she and her family were seen as "poor people" by the rest of the Afghan community in San Diego. This was one of the main reasons that Seema felt conflicted. Her low status among Afghans in the US created an identity conflict for her. She tried to compensate for this loss in social class by ensuring that she and her family wore expensive-looking clothes. They bought a TV the size of a wall and entertained frequently. On her shoulders was the burden of keeping up pretenses, which they really could not afford and which therefore added another burden.

Seema socialized mainly with other Afghan refugee women; therefore, maintaining class status was important for her since many of these women were aware of her illustrious family back in Afghanistan.

Kin expectations – an intergenerational understanding

Despite having close relatives in San Diego, kin support has been minimal for Seema because of family friction. She single-handedly mothers her four children, ranging from ages seventeen to five. The youngest child is a son. Finally having a son has been the crowning glory of her motherhood. After three daughters, she felt that not giving birth to a son was due to her inabilities. The eldest daughter, who is 17, is almost exclusively taking care of the third daughter, who is just around 10 years younger than she is, relieving her mom so that her mom can take care of the last child. Such co-mothering puts the burden on 17-year-old Amina. Amina gets up early like her mother does, helps Seema in getting the household ready for their respective activities such as getting the children ready for school, packs their lunches, goes to school, comes back, takes care of the third sister, squeezes in time to do her homework and goes to bed totally exhausted like her mother. The pressure on the oldest daughter is a common phenomenon in most Asian countries and is repeated in the diaspora. But here it gets layered with the daughter's ability to negotiate the outside world, and therefore, the expectations for her to help her mother are more intense and extensive, as is the case with Nadira.

For Seema, mothering is very tiring, and she would like to have some help. She misses her older sisters and aunts, who could have given her support if she were in Afghanistan. Seema's aging father also lives with her, and she performs the duties of a good daughter based on Afghan standards. Seema also took care of her aging aunt who died of cancer. She kept telling me that she has to be like a daughter to her aunt and take complete care of her because, she questioned, "who else will do it in this country?" Interestingly, the aunt had a son in Texas, but he also relied on Seema to take care of his mother because he saw caretaking as women's work. Care of older people in the community who substitute for close family members is also seen by these Afghan women as their duty, and they make sacrifices to be available for community members. The community then becomes the kin group. Seema and other Afghan women thus end up mothering not just their children but also older relatives.

Cultural adaptation – Afghanistan to Pakistan to the USA

In contrast to Shahnaz, Seema was exposed to a conservative form of Wahhabi Islam that was seeping into Pakistan. She and her family started following the Islamic dictates of the state and, most importantly, the Afghan refugee community. Seema came to the USA post-9/11 with two very young children and was suddenly confronted with a country about which she had no clue. She confronted a strong anti-Afghan and anti-Islam environment.

Seema does speak English and has become confident enough to negotiate the public world. Her attempt at partial assimilation was to adopt the Western dress of pants and shirts, though they are modest. She does not wear a headscarf except when she goes to Afghanistan or when she visits the mosque. As the girls have grown, there has been a shift in their dress. Now both the older girls wear pants or long skirts and sleeveless shirts but never tank-tops. When I talked to Seema about these changes, she proudly claimed that they need to adapt to Western culture and not stand out in school.

Simultaneously, Seema is proud that her children know their roots and also that they are good Muslims. Seema is very conscious of not letting the children stray from their Afghan roots. This is visible in the foods they eat, language they speak at home and their frequent calls to family members in Afghanistan. Despite their strained economic situation, they have prioritized going back to Kabul to meet family every few years. The biggest influence of Afghan culture in the home is Afghan TV. There is a channel out of Los Angeles which has 24 hours of Afghan TV shows and Indian Bollywood films. (The Indian film industry, located in the city of Bombay, is referred to as Bollywood, borrowing from Hollywood.) Bollywood films have for decades been and still are extremely popular in Asia and the Middle-East. Given this home environment, the children are constantly reminded of their roots.

Among recent refugees to the USA, especially those from camps in Pakistan, bringing children up as Islamic is seen as the most important responsibility of the mothers. Even though there is preaching of Islam as a unifying faith for all Muslims, cultural delineations are maintained. Recently, an Afghan mosque has been built in San Diego despite the existence of two large mosques in the city. Seema is very relieved by this because she sees Islam as the substitute for the extended family. Her children spend substantial amounts of time over the weekend in the mosque doing various activities as well as reading the Quran. The mosque does not act entirely as a religious institution; it is also seen as a community center. For Seema this has been a boon because now she has a larger Afghan community to help her acculturate her children. The mosque is also a contributor to Seema's creation of "home". It symbolizes for her the religiosity that she grew up with and further imbibed in Pakistan to become a better mother. This mosque is used mainly by recent immigrants as a community space to meet their people and to provide Islamic and cultural classes to Afghan-American children. It has also become the "watering hole" for new immigrants, who are provided services by the community.

Language is also a very important cultural tool for Seema. Speaking in English is strictly prohibited at home, as is watching English language television shows. An interesting phenomenon, though, is that Seema's son, the joy and pride of the family, is allowed to watch cartoons and play video games. As is the case with Shahnaz, Seema pampers her son. The birth of sons elevates the status of mothers within the Afghan communities in the USA as it does back home in Afghanistan. Seema's husband, too, though very liberal in many instances, proudly announced that they would not have any more children, reinforcing the prevalent and deeply entrenched patriarchal ideology of Afghans and other Muslim communities in the region.

A complicated "home"

I will now discuss briefly the concept of "home" (George, 1999; Khattak, 2002), not a new concept but relevant in a more nuanced interpretation to this discussion. As just observed, conflicts lead to disrupted communities and scattered families. Disrupted communities also mean that the common protective factors, particularly the family, may not be available. On a psychological level, loss of "home" for all refugees causes immense mental and physical distress. While discussing the "constant disruptions" Afghan refugees face in Pakistan, Khattak (2002) draws attention to the psychological aspects of displacement:

> While declaration of war is seen as an attack upon the greater home (state), attacks on actual homes in the context of armed conflict is not taken into consideration at all and does not even emerge as an issue of concern. The destruction of home and community has implications that go beyond the physical being of these places. These range from ideas of self, of identity, creativity, interpersonal relations and one's worldview. (Khattak 2002, p. 117)

Khattak continues by saying that the leaving of one's home during conflict compounds people's insecurity, especially women's, because it is layered with loss of home, loss of the familiar and entry into a new domain that is not comfortable and secure and that simultaneously is dangerous and anxiety ridden. Refugees also suffer the harsh realities of the loss of relatives and family members. Second, displaced women must remain flexible in the midst of great change, especially within the family unit. The "imagined" home is disrupted, and within the harsh realities of refugee camps, the home left behind due to fleeing seems a "golden" space – even though it's not perfect, it is one of security and predictability. Khattak (2002) concludes that

> For many women memory is an important coping mechanism. Memory serves to preserve their class and social identities, and also their national identities and association with their country as something beautiful. (Khattak 2002, p. 134)

A discussion about home is not just about land/country but about culture, kinship and family; it is about a sense of familiarity, language and food. The transportation of some of the above to a new location is still not seen as home, because of its fragmented nature and because it is a renegotiation of such institutions on Western terms. Displacements from home and country evoke a deep sense of loss and resentment as well as despondence. (Khattak, 2002)

In my interviews with Shahnaz, Seema and other Afghan mothers, I found a range of responses where any talk of mothering always led to a comparison of this process with either their own mothers or how it is different in the USA compared to in Afghanistan. Most of the responses also led to a discussion of "back home" for these women because not only was mothering about their taking care of their children, but it entailed and symbolized their struggles to find an Afghan identity as wife and mother in the diaspora.

An additional interesting phenomena not included in this chapter is the role of older mothers – older women who have received refugee status and have moved to the USA to live with their adult children. Anecdotally, and just from my observations, there seems to be a clear role reversal where these mothers become totally dependent on their adult children and are treated as children by their kids. Their dependency is entirely due to their inability to communicate with the public sphere, on the one hand, and with the new domestic situation they have arrived into, on the other. From how to use the washing machine to the lack of socializing with kin members and neighbors, this generation of mothers is handicapped, and this handicap is further reflected in their lack of status as mothers. They end up cooking authentic Afghan food, watching Afghan TV and babysitting their grandchildren. There is a severe generation gap with their grandchildren as well. These women are deprived of all their maternal role that earlier defined their status in society.

Conclusion

For Seema, the concept of home became much more complicated than it was for Shahnaz. The harsh reality of camps in Pakistan to the not so economically harsh but socially estranged society of the USA created its own problems for Seema. What became most complicated in this process of mothering for Seema was the concept of home. Is Afghanistan home for them, the place where they were born and spent their early formative years; Pakistan, which despite harsh conditions, protected them from the violence of Afghanistan; or the USA, which has provided them shelter from both the vulnerable and precarious situations of Pakistan and Afghanistan?

For women to deal with these dilemmas makes them less grounded, more vulnerable and definitely confused in the mothering process. This process is further complicated by the perceptions of their children who are born here or came here early enough to be socialized into Western society that they now call their home. Shahnaz expressed a forlornness when discussing the cultural gap between her and her children. This cultural gap redefined hierarchical relationships between parents and children and the expectations that parents would like to have from their children.

For Afghan mothers, the home becomes their cocoon through which they juggle the transmission of their cultural values to their children with their new found cultural expectations and exposures to a society that is different from theirs. For the mother who stays at home, this so-called Western culture is alien because of their non-exposure to it. But those who work outside the home, such as Seema, are also conflicted by the new scenario. As a refugee she is burdened with her responsibility to single-handedly maintain Afghan culture and to inculcate her children with Islamic values. Her dilemmas and convictions become just hers in a culture that neither understands them nor provides services for these needs. Simultaneously, the culture that she has been imported into is perceived by her as hostile because of her ethnicity and religion. Seema is constantly talking of returning home to Kabul. As pointed out by Khattak (2002), such sentiments are constantly expressed by refugees since such sentiments are "accompanied by various recollections and depictions of home representing peace and plenty as an important coping mechanism as there is hope that this memory can be revitalized sometime in the future" (p. 123). Unlike Seema, Shahnaz has reconciled seeing San Diego as her home. She says this has brought her a peace of mind.

What is very apparent from the differing experiences of Shahnaz and Seema is also the shift in USA's perception of Afghanistan and Islam. Shahnaz's family's migration to the USA coincided with a very different perspective of Islam and Muslims in general. In the 1980s Islam was not equated with terrorism, and Muslims were not seen as a threat to national security. Ironically, Afghans then were very welcome to the USA because they were fleeing occupation by the Soviets. Such was not the scenario post-9/11 when Seema immigrated to the USA.

The hybridized culture that emerges especially among Afghan children born and brought up in the USA is still alien and unsafe, nor is it comfortable to Afghan mothers. Thus, the trauma of war, remigrations and displacements challenge their identities and place in society, family and community. The constant renegotiation of their roles to be accepted by their families, their peers and non-Afghans is shouldered by mothers. It is reflected in their mothering patterns when they perceive their final mission as making sure their children remain culturally theirs at the same time that these children are also recognized and accepted by the larger US society as contributing members.

Afghan mothers carry the burden of being good Muslims; raising good Muslim Afghan children; and preserving the cultural integrity and honor of the family, clan and country – which is defined through the "purity" of women. For these women, individual identity is defined through only enhancement of a collective that is struggling in the diaspora to retain and respect its traditions. Their efforts define them despite the fact that the collective's gendered hierarchies appreciate masculinity and patriarchy above the strengths and value of women.

Although I am not a clinician, it doesn't take much effort to realize that these contradictions can create psychological conflicts in the women that may reflect interpersonal family conflicts. Not only gender but also generational conflicts may be present in more or less subtle ways between parents bent on preserving "home" traditions and their "American" daughters or sons, despite mothers' efforts to stress family unity and common customs. Some of these are present in all immigrant and refugee communities, others are uniquely relevant for Afghan families in the diaspora.

Service providers should inform themselves about this population to provide appropriate support to Afghan women and their families. As laid out in this chapter, Afghan mothers' dilemmas given their new location in San Diego, the new politico-social environment of the USA post 9/11, and the ongoing war in Afghanistan, creates a mind-set that distinguishes them from other refugees in the USA.

References

Espín, O.M. (1999). *Women crossing boundaries: A psychology of immigration and transformation of sexuality*. New York: Routledge.

Foner, N. (1997). The immigrant family: Cultural legacies and cultural changes. *International Migration Review, 31*(4), pp. 961–974.

George, R.M. (1999). *The politics of home, postcolonial relocations and twentieth century fiction*. Berkeley, CA: University of California Press.

Indra, D. (1999) Not a room of one's own. In D. Indra (Ed.), *Engendering forced migration* (pp. 1–22). New York: Berghahn Books.

Kallivayalil, D. (2004). Gender and cultural socialization in Indian immigrant families in the United States. *Feminism & Psychology, 14*(4), pp. 535–559.

Khattak, S.G. (2002) Violence and home: Afghan women's experience of displacement. In C. Calhoun, P. Price & A. Timmer (Eds.), *Understanding september 11*. New York: The New Press.

Martin, D.C. & Yankay J.E. (2014). Annual flow report: Refugees and asylees 2012. *Official Website of the United States Department of Homeland Security*. United States Office of Immigration, Apr. 2013. Retrieved from http://www.dhs.gov/sites/default/files/publications/ois_rfa_fr_2012.pdf (March 24, 2014).

Matsuoka, A. & Sorenson, J. (1999). Eritrean Canadian refugee households as sites of gender renegotiation. In D. Indra (Ed.), *Engendering forced migration* (pp. 218–240). New York: Berghahn Books.

McSpadden, L.A. (1999). Negotiating masculinity in the reconstruction of social place. In D. Indra (Ed.), *Engendering forced migration* (pp. 242–260). New York: Berghahn Books.

Mortland, C.A. (1994). Cambodian refugees and identity in the United States. In L.A. Camino & R.M. Krullfeld (Eds.), *Reconstructing lives, recapturing meaning* (pp. 5–27). Australia: Gordon and Breach Publishers.

United Nations High Commissioner for Refugees (UNHCR) 1951 Convention, art. 1A (2), 1967 Protocol. Retrieved from http://www.unhcr.org/3b66c2aa10.html (March 20, 2014).

Part III

Violence, Resistance and Resilience

Introduction to Part III

Oliva M. Espín and Andrea L. Dottolo

Violence is a constant fact in the lives of women in all societies, including mainstream US society. Although US-born women face obstacles similar to those of immigrant women in situations of domestic violence, sexual violence, sexual assault and other forms of victimization and harassment, there are several cultural barriers and specific circumstances that intensify these problems for immigrants.

Evolving family power dynamics that occur after relocation (Darvishpour, 2002) may result in increased incidences of domestic violence. Although statistics are not fully reliable, it appears that women immigrants are more likely to experience domestic violence than US-born women. It seems that immigrant-specific factors exacerbate the already vulnerable position of immigrant women in situations of domestic violence (Menjivar & Salcido, 2002). In a study with 280 immigrant Latina women ages 14–56, Hass, Dutton and Orloff (2000) found that 136 women reported experiences of physical abuse by their intimate partner whereas 33 women reported experiences of psychological abuse. Women who have refugee status experience similarly high incidences of interpersonal violence. Concerns and fears such as visa status or fear of deportation for self or the spouse may prevent immigrant and refugee women from asking for help (Abraham, 2000). Immigrant women may be unaware that partner abuse is against the law in the US because it is legal in many other countries. Moreover, these women may be wary of requesting police services in the US due to law enforcement inefficiencies in their host countries.

The US federal government has made provisions for women and their children who are abused with the passage of the recently reauthorized Violence Against Women Act of 1994 and the Battered Immigrant Women Protection Act of 1999, the goal of which has been to remove immigration laws as barriers that had kept battered immigrant women and children locked in abusive relationships. Over more than two decades, this law has saved countless lives. The original law created the National Domestic Violence Hotline and authorized federal funding for battered women's shelters. New laws provide a temporary visa status and protections against deportation to undocumented immigrant women and children who are experiencing domestic violence in their homes, but multiple barriers continue to exist for women who find themselves in these situations.

Especially difficult is the situation of women who are undocumented immigrants and who experience domestic violence while in the US. Women who reside in the US without documents are least likely to seek help and protection from domestic violence. Moreover, abusers often use their partners' immigration status as a tool of control. They may subject her to constant threats of deportation, threatening to report her to the authorities to deport her and keep the children. They may destroy important documents such as a passport or identification cards, or they may make constant derogatory comments about her or her culture and race. It is important to note here that the abusive partners of immigrant women may be US-born white men.

Also, immigrant women are often more likely to stay in relationships due to the varying social conventions of their countries of origin. This creates the fear not only that they may be deported if they report their batterer to the police, but that they may also have to face family members as a shunned divorcée.

Finally, language serves as the most obvious barrier that prevents many immigrant women from taking legal action against their abusers. Not only may a woman have difficulty conveying the physical and mental abuse perpetrated against her by her batterer, but she may also be isolated from the legal and social services available to her due to a language barrier.

Yet another oppressive factor comes from people eager to be "culturally sensitive". Frequently, when outsiders to a given cultural group are trying to be "sensitive", they attribute behaviors that would be unacceptable in members of the majority culture as being part of "the culture" of this group. This perspective can distort efforts to enhance cultural awareness into dangerously destructive practices for women. This may be even more destructive because they are carried out with such good intentions. Conscientious US professionals often believe that wife abuse is "part of these people's culture". Therefore, it should be tolerated. Women and children are seen as the suitable recipients for oppressed men's displaced anger. Obviously, wife abuse is *also* part of mainstream

US culture, and as helping professionals or educators, we would not dream of advocating tolerance for it in mainstream US culture.

A common legal issue contributing to women's struggle to gain protection from violence is that "culture" is used as a defense against the victims of domestic violence (Renteln, 2002).

Even though the displacement of anger onto women and children among men who feel deeply frustrated in their efforts to obtain recognition from mainstream society can be understood by professionals, acceptance of violence against women as an outlet for their frustration or as a legitimate reaction to their own oppression is misogynistic and immoral. It is unacceptable to justify it in the name of "culture". And it should *not* be advocated as a manifestation of cultural sensitivity.

Conversely, much is made about the incidence of male dominance in immigrant cultures by individuals in the host culture. Immigrant individuals are often portrayed as criminal, poor, violent, desperate to live in the US and uneducated, regardless of what their actual lives and level of education might be. However, it is important to remember that any expression of male dominance among immigrants is nothing but the specific culture's version of the myth of male superiority that exists in most cultures, including mainstream American culture. Although many immigrants still subscribe to the traditional ideas of male superiority and its consequent forms of expression, many reject it outright. Let us remember that there are many immigrant women who are actively involved in the feminist movement and who are unwilling to submit to the authority of male relatives. Indeed, some of the most effective programs to prevent violence in immigrant communities have been developed by groups organized by immigrant women themselves (e.g. Abraham, 2006; Perilla, 1999).

The chapters is this section, address issues of violence against women that include domestic violence, but are not limited to it.

In the first chapter in this section, Kallivayalil presents a feminist and trauma-informed psychological perspective on the complex challenges faced by women who are victims of *politically motivated violence* and who are seeking political asylum. Material from clinical cases is utilized to highlight ways of working with this population in a culturally informed manner.

Tummala-Narra, Satiani and Patel provide a gender analysis of aspects of South Asian culture, including religion, spirituality, sexuality and family structures and dynamics. Using a feminist psychodynamic lens, this chapter aims to extend existing clinical understandings of social and contextual issues that contribute to the problem of *sexual violence*, the recovery process and empowerment within South Asian communities.

Quintero, Cerezo, Morales and Rothman discuss a community-based qualitative research project that investigated the lived experiences and consequent mental health needs of transgender women who migrated to the US from Latin America. Perhaps no other population suffers as intensely under the effects patriarchal domination and *violence as transgender women*. Transgendered women who are immigrants face yet additional layers of struggle, prejudice and discrimination. Because cultures so often conflate gender identity with sexuality, the confusion, hostility and resentment that transgendered immigrant women meet in the host country is compounded by many other factors related to immigration, including language, social class, and so on. They suffer additionally from racism and the effects of legal constraints imposed on all migrants. Further, gay and lesbian communities in the host country are not necessarily welcoming of transgender women for a variety of political and social reasons. In this study, experiences of violence, addiction and other traumatic events are present, but so are stories of mutual support and resilience described by the women interviewed. The case study of Erika illustrates these issues and inspires an analysis of the specific circumstances of this group of immigrants. Recommendations for professionals' development to better understand and intervene in the lives of transgender immigrant Latinas are included.

Serrata, Macias, Rosales, Rodriguez and Perilla present the discoveries of a study of a group of Latina immigrant survivors of *domestic violence* who completed a leadership development program (the *Líderes* program) in the Southeastern United States. Phenomenological data analysis is used to identify prevalent themes. The authors discuss the importance of acknowledging histories of violence and of the supportive environment provided by the program, the personal and observed discovery of natural leadership abilities and a collective commitment to creating community change. The program described in this chapter is an illustration of the role that resistance and resilience in the lives of immigrant women who are survivors of violence in its many forms.

Together, these four chapters provide concrete clinical examples and strategies to deal with the complex and multifaceted worldwide problem of violence against women.

References

Abraham, M. (2000). *Speaking the unspeakable: Marital violence among South Asian immigrants in the United States*. New Brunswick, NJ: Rutgers University Press.

Abraham, M, (2006). Mode; minority and marital violence: South Asian immigrants in the United States. In R. Mahalingam (Ed.), *Cultural psychology of immigrants* (pp. 197–216). Mahwah, NJ: Lawrence Erlbaum Associates Inc, Publishers.

Darvishpour, M. (2002). Immigrant women challenge the role of men: How the changing power relationship within Iranian families in Sweden intensifies family conflicts after immigration. *Journal of Comparative Family Studies, 33*, pp. 270–296.

Hass, G.A., Dutton, M.A. & Orloff, L.E. (2000). Lifetime prevalence of violence against Latina immigrants: Legal and policy implications. *International Review of Victimology, 7*, pp. 93–113.

Menjivar, C. & Salcido, O. (2002). Immigrant women and domestic violence: Common experiences in different countries. *Gender & Society, 16*(6), pp. 898–920.

Perilla, J.L. (1999). Domestic violence as a human rights issue: The case of immigrant Latinos. *Hispanic Journal of Behavioral Sciences, 21*, pp. 107–133.

Renteln, A.D. (2002). In defense of culture in the courtroom. In R.A. Shweder, M. Minow & H.R. Markus (Eds.), *Engaging cultural differences: The multicultural challenge in liberal democracies* (pp. 194–215). New York: Russell Sage.

6

Women, Political Violence and Migration: Feminist and Clinical Issues in the Asylum Process

Diya Kallivayalil

A woman flees domestic violence, state persecution and the threat of forced sterilization only to be ensnared as a victim of sex trafficking. A teenage girl escapes her village so that she will not be forced to endure genital mutilation only to face discrimination and unemployment in her new country. A woman is raped, mutilated and displaced by warring militias and forced to leave her children behind as she escapes across the border. These women and their stories are familiar to clinicians working with women survivors of political violence. Unfortunately, their experiences are also illustrative of the structures of political violence that place women – as women – at risk. This paper considers the nature and scope of politically motivated violence against women, offers a feminist critique of the current political asylum system, discusses the complex psychological challenges faced by women who survive these experiences and offers recommendations to clinicians working with this population.

Politically motivated violence and political asylum: The scope of the problem

The number of people and communities impacted by politically motivated violence, persecution and torture has reached truly alarming proportions (Drozdek & Wilson, 2004). Recent estimates put the number of current refugees at 15.2 million people (UNHCR, 2011) and indicate that over half a million survivors of political violence live in the United States (Office of Refugee Resettlement, 2006). Asylum claims in industrialized countries have increased dramatically, with the majority of these claimants living in long-term displacement from their countries of origin (UNCHR, 2011). The United States currently receives more asylum claims than any other industrialized nation. These statistics indicate that political violence is a pressing political, public health, human rights and pressing political, public health and human rights issue.

For the purposes of this discussion, the term *politically motivated violence* refers to inter-nation war; intra-state political, ethnic, secessionist or religious conflicts; human rights violations within countries, such as arrests, acts of torture and disappearances for political activity that opposes the government or ruling party; or state laws or a lack of legal protection that violates basic rights of individuals as members of social groups (e.g. women, gay and lesbian individuals, members of certain religious groups, members of minority ethnic groups, etc.) and therefore requires them to flee their country of origin and seek asylum or become refugees elsewhere (Kallivayalil, 2013). The United Nations 1951 Refugee Convention definition for seeking protection (based on the Geneva Convention) is as follows: "owing to a well-founded fear of being persecuted for reasons of race, religion, nationality, political opinion or membership in a particular social group" – the definition of "social group" may include gender but harm based on gender is not currently specified as a legal basis for protection and political asylum (Kea & Robert-Holmes, 2013; Middleton, 2010; Sircar, 2006). The implications of this will be discussed later in this paper.

Women's exposure to politically motivated violence and symbolic use as "spoils" of conflict

As numerous feminist scholars have shown, women are disproportionally impacted by politically motivated violence (Bhabha, 2007; Clifford, 2008; Kelly, 1993; Stark, 1996; Ziegler & Stewart, 2009). During periods of political conflict, women are made to suffer, in addition to other forms of violence, loss, displacement – pervasive sexual violence. This includes organized rapes, torture, physical mutilation (including cutting off of lips and ears, blinding, amputations of limbs, and mutilations of genitalia), forced kidnappings and forced prostitution (Clifford, 2008). These descriptions merely scratch the surface of the atrocities women suffer during political conflict and instability. Mass rapes during conflict are increasingly employed to terrorize communities, destroy families and shred the prevailing social fabric (Clifford, 2008, p. 4). In the current conflict in Eastern Democratic Republic of Congo (DRC), for example, family members are often forced to watch and participate in the rape and mutilation of their women family members (Kelly et al., 2011). Women and their families are further victimized by a pervasive atmosphere of fear of sexual violence, which has profound psychological consequences. The violation of women is considered a tactic of war in which the goal is to humiliate and demean men by dehumanizing and destroying women, who are seen as their assets and property (Clifford, 2008; Bartells et al., 2010; Kelly et al., 2011; Stark, 1996; Bhabha & Shutter, 1994). Families and communities are thus ripped apart by shame, incest and betrayal. "When rape is employed instead of a bullet, the weapon continues to wield its power beyond the primary victim, while the battlefield may be the body, the target is civil society" (Clifford, 2008).

Adding insult to very real injury, women who flee their countries and communities are often faced with considerable difficulty when they bring political asylum claims forward and cite their experiences of rape and other forms of sexual violence. Too often, "Authorities tend to view rape as an apolitical, 'personalized' act of random violence against individuals, rather than a component of a political strategy against persecuted groups. Rape is still frequently viewed as a crime of honor rather than a crime of torture" (Bhabha & Shutter, 1994, p. 253) – a view that also explains why these women are often ostracized by their communities and families. They are seen as having dishonored their husbands and families and are frequently divorced and prevented from remarrying. They and their children then live at increased risk for further victimization. Indeed, there is increasing evidence that the fear and stigma of rape and women's fear of returning to their communities lies behind larger numbers of asylum claims in Western countries (Amiss & Neale, 2006). Women are also at risk in post-conflict settings which are frequently characterized by "volatility, weak, social institutions and widespread impunity" (Kelly et al., 2011, p. 21; Ahmed-Ghosh, 2006; Bhabha, 2007). Their children who may have been born of rape are shunned and often forced to live as orphans and in foster homes. The psychological impact of assault and social rejection for women and their children is profound. In sum, their identities as women are used as the basis for attack and violation; at the same time, these women are held responsible by their own communities for not upholding their virtue as women and mothers and as symbols of the country's or community's honor. Zavos (2010) argues, "If the women of a country are violated, it is assumed that the whole country is violated as well, as if, in the end, women were the bearers of a country's territorial, moral and social integrity and wholeness" (p. 23).

Scholars have also documented the impact of pervasive sexual violence during military or intra-state conflict and political instability on women's economic survival (e.g. Bartels et al., 2010; Clifford, 2008; Stark, 1996). As routes to markets and jobs are disrupted, the fear of sexual violence prevents women from farming, going out into the fields to collect water or going to school or work (Kelly et al., 2011; Clifford, 2008). Women who have been assaulted have numerous medical and psychiatric needs, which are not addressed in conflict settings. Hunger, malnutrition and disease are frequent consequences of the threat of sexual violence (Kelly at al., 2011). They are often forced to migrate or cross borders during conflict only to face rape, torture and other forms of violence (including witnessing the rape and torture of others) as they cross the border by soldiers, para-military organizations or men who are "travel brokers" and are assisting them with border crossing (Paris, 2008). As refugees or internally displaced persons, they may be required to live for years in long-term settlement camps and may be exposed to further sexual violence (UNHCR, 2011).

Gender-specific persecution

Women also seek asylum due to gender-specific persecution, which is considered harm *particular* to their gender (Kea & Robert-Holmes, 2013). This persecution includes rape, domestic violence, female genital mutilation (FGM), child marriage, bride burning, forced marriage (such as being forced to marry one's rapist to avoid punishment for sex outside marriage) and forced sterilization and abortion (Bhabha, 2004; Stark, 1996; Sircar, 2006). Women are also impacted by state level decrees and social mores dictating gender-specific behavior, which frequently relate to women's rights to control their bodies or reproductive organs (Bhabha, 1996). There has been an increasing acknowledgment that human rights violations such as rape, FGM and domestic violence can form the basis for a claim of persecution (Middleton, 2010). However, the recognition that gender-specific forms of protest (such as a refusal to wear the veil) or political activity could also constitute such a claim has been slower and more complex to evolve; political activity is generally seen as the domain of men while issues affecting women are seen as private or personal, even when the consequences for this activity are equally severe (Bhabha and Shutter, 1994; Kelly, 1993; Middleton, 2010; Sircar, 2006). There is also a prevailing view that women (and children) migrate as dependents of men (Kea & Robert-Holmes, 2013; Thapan, 2006). Thus women who migrate outside the context of the family unit (such as women fleeing domestic violence) face additional legal burdens.

Feminist critiques of the political asylum apparatus

Despite the fact that the political asylum system was instituted in part to provide safety, legal redress and options for those that have been persecuted, the process of obtaining political asylum and the dominant human rights narrative around this issue is problematic, and it has been critiqued by feminist scholars (Bhaba & Shutter, 1994). Ahmed-Ghosh (2006) cautions that the discourse of human rights is contested and potentially complicit in its own version of hegemony – "using the western status of women as a standard of measure" of the political standing of a nation or what Herzfeld (2007) calls "the global hierarchy of value" of countries. This has implications for political will and political and social action. Further, there has been a tension in asylum and refugee cases between recognizing the systematic oppression of a group and recognizing the systematic targeting of individuals (Sircar, 2006). Sircar (2006) argues, "At times fear of treading on state sovereignty has led advocates for cultural relativism to warn against interpreting traditional and historically sanctioned practices as persecution" (p. 259). This has particular implications for issues like domestic violence and FGM, where there has been a tendency to ascribe these acts to "private actors" or cultural norms.

These issues are also closely connected to the mechanisms of applying for political asylum and the inherent notions of "women as victims" or "women as problem" (Crawford & Maracek, 1989; Ghorashi, 2005; Hadjukowski-Ahmed, 2007; Sircar, 2006; Thapan, 2006). Historically, perspectives of migration have been limited because explanations have either implicated individuals' "rational" decisions or economic determinism. Both are of limited value and need to be deconstructed when it comes to understanding forced migration and political violence (Thapan, 2006). Kea & Robert-Holmes (2013) critique the current political asylum system by arguing that female asylum seekers are forced to be partly complicit in the creation of a victim identity for the purposes of asylum. The representation of women as victims is problematic, particularly because the corollary of this is that these women need saving by the more enlightened West (Kea & Robert-Holmes, 2013). However, it is also problematic to discount this representation and instead look simply for women's resistance and resiliency, which is also ultimately essentializing and simplistic. Palmary (2010) argues, "There has been a tension between an overstated homogenous approach to power that allows no space for agency and an overly romanticized understanding of resistance that both sees it everywhere and celebrates its effects too easily" (p. 56). It is important to move beyond the dichotomy of women as victims or of women as victors or survivors to a deeper analysis of power and gender relations. As Abu-Lughod argues (1990), "We could continue to look for and consider nontrivial all sorts of resistance, but instead of taking these as signs of human freedom we will use them strategically to tell us more about forms of power and how people are caught up in them" (p. 42). Further, although there is a growing human rights literature on the challenges facing women in conflict settings, there is less work representing women's own voices and calling attention to racial, ethnic and class differences among women, which may play a large role in their experiences of conflict and post-conflict settings (Ahmed-Ghosh, 2006; Kea & Robert-Holmes, 2013). Along these lines, this paper calls attention to the systemic and discursive problems of political violence and the political asylum apparatus. It also examines the difficulties of individual women and their day to day struggles as legitimate objects of study in their own right.

Psychological impact of political violence

The psychological impact of various forms of political violence and torture is increasingly well documented and points to post-traumatic stress disorder (PTSD), depression and other mood disorders, complicated grief and substance abuse problems (see, for example, Eisenman, Keller & Kim, 2000; Amowitz, Reis, Lyons, Vann, Mansaray, Akinsulure-Smith, Taylor & Iacopino, 2002; Hooberman, Rosenfel, Lhewa, Rasmussen & Keller, 2007). These issues are in addition to the physical effects of torture and abuse, which may include broken bones, joint and muscle pain, headaches, burns and scars, neurological damage such as hearing or vision loss, reproductive problems, dizziness and loss of sensation (Iacopino et al., 1999). There is also evidence that these survivors are also at higher risk for more prolonged medical illnesses.

Women's lived and internal experience of political violence: Key psychological issues

Given that the psychological impact of political violence has been well documented, this paper asks: what psychological issues are important to women who have been exposed to this violence? What concerns do they raise in clinical treatment, and what issues should clinicians working with this population pay attention to? How can their experience inform feminist understanding and theory, public policy initiatives and meaningful approaches to clinical care?

Understanding women's lived, internal and psychological experiences of political violence requires not only a documentation of symptoms but also attention to the following intersecting issues, which are of central importance to a feminist analysis: (a) the impact on women's working lives and their entry/retreat into the workplace and (b) decisions made during migration that impact parenting and separation from children. Survivors of political violence are a population with particular needs under the umbrella of trauma-informed work, and these issues illustrate just some of their dilemmas and experiences. They are discussed below with brief case material to illustrate their complexity. As the cases will demonstrate, these issues are deeply interconnected.

Impact on women's working lives

Forced migration as a result of war, political violence and gender-specific persecution has long-lasting implications for women's working lives. The jobs they held prior to migration are obviously disrupted and lost. When these women arrive in their new country, they continue to face gender-related dilemmas in relation to work. Some women did not work in their home country for various reasons, including class status, relative wealth or gendered prohibitions against women's working outside the home. They may find themselves in a new and unfamiliar role, such as the sole provider for their children and themselves (Kelly et al., 2011; Burnett & Peel, 2001). Further, it could take these women months or years to receive their work authorization once they enter the asylum process. They are faced with the choice of working "under the table" to support themselves and any family members they have come with and/or to send money back to family members, including children for whom they are often the sole provider as war and political conflict may have ravaged the children's local economy. For the women who chose this route, this often means an engagement with their national or ethnic community in their new country to find work opportunities, which they may feel ambivalent about, given the political and ethnic conflicts and realities in their home country (Kallivayalil, 2013). It also places them in a position of being beholden to their employers, many of whom abuse this power by forcing women to work in sweatshops or as domestic servants or by sexually abusing them. These employers also frequently take away women's passports and threaten to expose them to the authorities who will deport them, often telling women that their families back home will be targeted and harmed.

As an illustration of some of these issues, my client Mai left China fleeing domestic violence, persecution and the threat of forced sterilization for violating the "one-child policy". She was promised housing and a job in a shop by a friend of the family once she arrived in the United States. She left behind her five-year-old daughter in the hope that they would one day be reunited. The "job in a shop" turned out to be a brothel hidden behind a dry cleaning store. About a week after she began working, she was told that she would have to engage sexually with "clients" and was told that if she did not, her employer would report her to the police and she would be sent back to China. Her boss began to sexually harass her, and he raped her on several occasions. She eventually escaped with the help of law enforcement, but soon after, an old co-worker called her to warn her that her old boss knew that she was working with police and asked this co-worker to deliver a message: We will find your daughter in China and harm her. Mai said, "I was so scared and also ashamed. I did not know who I can turn to for help. I think about ending my life often, but I would never do it. I have my daughter. I worry about her safety, and this bothers me more than anything that I have been through."

Dilemmas continue even when women are permitted to work legally. Some women may have had lives of relative privilege in their home country and may have worked in high status jobs. When they are forced to migrate for political reasons, they may find themselves unable to find jobs of similar stature or pay and find themselves working in "low-status jobs" – for example as food industry workers, cleaners or nurses' aides – which they find deeply injuring, in addition to the fact that their wage makes it very difficult to live or accumulate money. My client Afia was targeted due to her husband's political activity in a Central African country. She had a master's degree and worked in a bank in her home country. She said, "I feel very depressed. I had a good job back home and my own office, and here the only options for me are to work in a place like Dunkin' Donuts. And then people say I should be grateful for this job. I feel very discouraged when I hear from friends from back home and get news that their careers are progressing. I have been avoiding looking on Facebook and calling home."

Conversely, other women have grown up in relentless poverty and/or in societies where they are not permitted to work for money. They therefore often see the ability to work as an opportunity and enjoy and take pride in working. My client Patricia, who fled civil conflict in El Salvador, found work as a cake decorator in a high-end bakery. She often discussed how proud she felt when her boss and customers complimented her delicate and detail-oriented work. However, some women say that although they see opportunities, they feel they continue to find jobs in gendered professions. My client Barbara, who was a victim of trafficking, was arrested in a raid of her brothel. The charges against her were eventually dropped, but the arrest remained on her criminal record. She said, "I applied for a job in a school, and I didn't get it. I think it's because of my background check. So now I am working as a waitress in a restaurant, and I am worried that this work is the only work I will be able to do. The hours are very long. I feel so disappointed." There is also a subset of women seen in clinical practice who are too impaired by their psychiatric symptoms and physical injuries to work, but their numbers are not known.

A discussion of women and work within the context of political violence would be incomplete without a deeper examination of sexual and domestic trafficking. The United States receives the second largest number of sex trafficking victims in the world (Mizus et al., 2003; see Reid, 2010 for a systemic review of human trafficking). There is a close link between political conflict and instability and sex trafficking. Study after study shows how community-level and individual factors intersect to increase vulnerability to sex trafficking: on the community-level, political instability, ethnic conflict, forced migration, gender discrimination, violence against women, armed conflict, government and police corruption (to name a few); and on the individual level, family violence, daughters sold by their families, childhood sexual abuse, abandonment, poverty, homelessness and pressure to support the family economically (see, for example, Acharya, 2009; Clawson, et al., 2009; Deer, 2010; Schauer & Wheaton, 2006). Economic arguments have long been primary in understanding both the mechanisms and the market for sex trafficking, but there are also feminist theories underscoring patriarchy and patriarchal practices in many countries and societies that support and allow the trafficking of women and girls (Farr, 2005; Gozdziak & Bump, 2008).

Neuwirth articulates the psychological relationship between sexual violence and sex trafficking as follows: "Brutalized by rape and violence, passed from trafficker to trafficker, and sold again and again in prostitution, these women are often broken by the force that has been used against them- it would be a grave injustice to mistake their submission for consent. It is not consent, but it makes force, fraud and coercion very difficult to prove" (2008, p. 3). Women who have been victims of sex trafficking following experiences of political violence feel that all aspects of their volition and sense of self have been destroyed. They feel deeply ashamed of their so-called participation in sex work because there is the perception that they gained economically from their participation when in fact they are most often isolated, threatened and terrorized. An all too common story is that women attempting to escape their home countries are promised jobs by people who are in the United States or by those in their own countries, only to arrive here and find themselves forced into prostitution or domestic servitude. Their passports are often taken away, and they are told that any notification of authorities will result in immediate deportation and a return to the violence they have tried to escape (Kallivayalil, 2013).

My patient Wei, who was targeted and tortured for her political activity in China, thought she would be working at a restaurant in Iowa. Her boss abused her physically and sexually and told her that if she went to the authorities, she would be deported back to China due to her lack of immigration status. She was then sent to work in a "massage parlor" owned by her boss, which turned out to be a brothel. Her boss was also abusive of other employees, so she felt that because others tolerated this abuse, there was no hope for her if she were to go to the police. She discussed the isolation she experienced: "I wanted to call the police, but I didn't even know where I was. I was so isolated. I had no connection to outside world. I had no idea what rights were mine. I spoke very little English. The customers also beat and raped me. I still have a lot of back pain from this. My bosses told me that if I disobeyed them, they would deport me and hurt my family in China. I also felt like I was a criminal because of my work in the brothel. I felt no one would believe me and fight for me, and I was so scared to go to prison or be sent back to China. When the police came to close down the massage parlor, they treated me well. They told me for the first time that I am not a criminal; I am the victim."

Migration factors that impact parenting and separation from children

Women who migrate for economic reasons (simplistic as this idea may be) often leave their children in their home countries with the plan that they will send back money to support them and/or their aging parents and the hope that they will be reunited with their children at some point (Paris, 2008). These children are most often cared for by extended relatives, often a grandparent (Hondagneu-Sotelo & Avila, 1997; Tummala-Narra, 2004). These women explain their decisions (to their children and themselves) that they are providing a better life for their children and that the separation will be temporary. Despite this hope, research indicates that many of these separations often last for more than ten years (Paris, 2008). These painful and prolonged separations place women at risk for depression, anxiety, psychosomatic problems and deep isolation.

These dynamics are multiplied tenfold for women who are fleeing political violence and are in fact forced migrants. There is generally no time to plan, little time to say goodbye and little time to explain decisions. Some women leave their children in their home countries without clear assurances of who will care for them, some are separated from their children during conflict and some have to make the decision to take some of their children with them and leave some behind. Some are forced to leave children behind with an abusive family member, often the child's father. What compounds the internal pain and horror of these decisions is that there is often no known plan or concrete hope for reunification. Women feel very deeply the judgment of others (imagined and real) for their decisions and wrestle with profound survivor's guilt. They also worry that their children and other family members will be harmed or retaliated against as punishment for their escape. Women who are in the process of applying for political asylum and thus in some ways known to the United States Immigration Service also live in fear of being separated from their children born in the US if they are sent back to their home countries (Paris, 2008). They also worry that children left behind are jealous or resentful of the children born in the United States. Women express the pain of these dynamics in poignant ways. My client Rosa, who fled a Central American country, left her son behind with an aunt. She has since had a daughter here. She said, "A bad day for me is when I cannot forget the guilt and worry. I sometimes try not to talk to my son because I feel helpless and he is angry with me. He thinks I have forgotten him and made a new family here. But then I feel guilty for not calling him." Another client, Akua, fled a North African country when civil war broke out. Her army officer husband was targeted for being on the wrong side of the conflict and was killed. Her 20-year-old son was not allowed to leave and thus remained behind in a camp: "I can never be happy here. Before the war, we had everything: a house, a car; we took holidays. Now everything is lost. If my son were with me, I would not care about anything, I'd just be happy: not the case, not getting asylum, not food, nothing."

Another client, Portia, witnessed a group of armed terrorists attacking a neighborhood and killing children, and she became a target for this group. They went to her home searching for her. The police were complicit in this group's activities. Her husband was killed, and she fled on foot to another country and eventually came to the United States, but she left her two young sons behind with her mother. She said, "I felt pressured from all sides. I could not trust the police. I think about ending my life, but it's completely against my religion. We would not want to lose in this life and the next one. When I talk to my sons, they don't complain. But they always ask when we will be together. And I tell them I am trying, but I feel so much pain when I hear this."

Clinical issues and recommendations for care

This section addresses some factors that may distinguish survivors of political trauma from other survivors of interpersonal violence (largely family violence) in the clinical realm. It may be useful for clinicians who work with this population to consider the impact of these issues on the clinical presentation.

a. *Systems issues*: Although trauma survivors are in general often faced with a myriad of interactions with systems (police, courts, and child protection services), women asylees and refugees face particular systemic obstacles that impact their symptoms and access to care and supports. These obstacles are often determined by where they are in their process for asylum and the specificities of each case. Generally speaking, the women who present for psychiatric care are in one of two situations: 1) they are in the process of applying for asylum and are in the position of having to prove to the legal system the legitimacy of their claim of persecution and their fear of future harm; or 2) they have entered the country without documents and are not planning to apply for asylum for various reasons but are seeking care because their symptoms are of such severity that they are struggling to function, work and/or parent. In either situation, these women are facing significant systemic barriers. As discussed earlier, they are not legally authorized to work (although some receive work authorization at some point in their application for asylum which can take months or years). As a result, they have very little or no money and lack access to the most basic resources, such as food, housing and transportation (Smith & Keller, 2007). Many have fled their countries of origin with little more than the clothes on their backs. Some of them are being temporarily supported or housed by members of their communities or religious organizations, but they are well aware that this support is limited or temporary. They frequently face language barriers as well (Burnett & Peel, 2001; Sircar, 2006). Some of them or their family members have been detained at the border or have been picked up by police or immigration officials and already have a record in the criminal system. In addition, they often face racism and discrimination as immigrants to their new country (Smith & Keller, 2007). Their lack of legal status and fear of authorities place them in a vulnerable position with regard to other forms of abuse – domestic violence, police brutality, forced domestic labor and abuse by the families or religious and/or community groups that have taken them in. These pressing issues are at the forefront for them when they present for psychiatric care. Clinicians may feel out of their depth in dealing with clients with such a large range of psychological and systemic needs and feel unsure about how to be most helpful to them.

b. Complicating this picture further is these women's frequent and understandable *lack of experience with psychiatric treatment*. Many have been referred by their attorneys or medical providers and are unsure what therapy or psychiatric medications entail. They often feel fearful of authority figures because in their home countries government officials, the police, armed forces and health professionals have frequently been involved in abuses of power and political repression (Kallivayalil, 2013; Smith & Keller, 2007). They are unsure what to say about their histories and symptoms because they often have histories of having undergone interrogation and are fearful of disclosing information they fear will be used against them or those they love. The concept of therapeutic confidentiality is looked upon with suspicion and mistrust. Women who are filing an asylum claim based on domestic or family violence or enforced social custom fear that their husbands or families will be notified of their whereabouts. All of these factors make them extremely wary of the therapeutic process. It is therefore crucial that clinicians inquire about and openly discuss issues regarding trust and confidentiality with their clients.
c. *Women's exposure to patriarchy and family violence, often prior or coinciding with their experiences in war or conflict*

There is now a substantial literature demonstrating how a history of childhood violence (physical, sexual, and emotional) can lead to significant long-term psychosocial impairment, places people at risk for further victimization and also impairs these same survivors' ability to cope with and mitigate the effects of such abuse (for a summary, see Courtois & Ford, 2008). Women who apply for political asylum often report having experienced childhood sexual and physical abuse and domestic violence long before they were impacted by the violence that forced them to become asylees or refugees (Radan, 2007). Recent research demonstrates that in Eastern DRC, an area that has become synonymous with organized militia-perpetrated rape, a substantial portion of women also report sexual abuse in their homes (Peterman et al., 2011). These early adversities have implications for these women's psychological vulnerabilities and their clinical care. There is a tendency among professionals to minimize or neglect to inquire about early abuse when the details of the political violence are so horrific and at the forefront (e.g. mass rape, mutilation, and genocide). However, when these women present in therapy, they are often very eager to discuss their childhood experiences and how those experiences may contribute to their current difficulties and also their sense of self. It is important not to neglect these discussions, as they are crucial components of building a treatment alliance and also shed light onto both clients' difficulties and their resilience, which can be mobilized in treatment. My client Rosa talked at length about growing up in abject poverty in her home country and being forced to steal to get food for the family: "I was made to go to the market and try to sell tortillas when I was seven years old. If we didn't sell enough, my father would tell me to steal. I never went to school. My brothers did not have to do that; they went to school. I sometimes understand why my mother allowed this, but sometimes I think I was not given love by her; she treated me as a servant because I was a girl. When I tried to come home when I tried to leave my husband, she asked me to leave. That, I can't understand. Then everything happened with the juntas. I feel my life has no purpose; I was put on earth simply to suffer. But I believe in God, and this helps me."

My client Sylvia grew up in a very poor home and was severely physically abused by her mother. She was then forced her to support the family through prostitution: "I was the oldest, and she told me, 'do you want your brothers and sisters to starve?'" Sylvia eventually fled on foot to the United States from Central America when she was targeted by local militia and kidnapped and raped. She said, "I feel upset all the time. My head feels heavy all the time. I feel my hands are not strong. The things with my family bother me more than anything. I had a difficult life in my country, but the things that have happened to me here have been so difficult as well. Sometimes I feel life is not worth living."

Conclusions and treatment recommendations

There are numerous ways for us to increase our attunement to and to be advocates for our clients who have survived politically motivated violence. Given our clients' frequent lack of experience with psychological treatment, it is important to set time aside to fully explain the role of psychotherapy and address issues of confidentiality, particularly if we are also writing a psychological affidavit to the court to support their asylum claim. It is also important for us to be informed about the asylum process and our clients' home countries, while also focusing on helping our clients negotiate their day to day existence in a new country. Assessing so-called pre-trauma experiences, such as family history and prior work experiences, is crucial to our understanding shifts in their internal world and their sense of self that can be addressed in treatment. We should also attend to issues of parenting and loss, their wishes and fantasies about reunification and their future. Finally, and more broadly, we should strive to avoid the disempowering view of seeing our patients as simply victims or to avoid romanticizing their endurance; rather, we should strive to understand the systems that touch their lives, before and after their migration and to support and affirm their multiple identities and daily struggles.

References

Acharya, A. (2009). The dynamic of internal displacement, forced migration and vulnerability to trafficking in Mexico. *Journal of Human Ecology, 27*(3), pp. 161–170.

Abu-Lughod, L. (1990). The romance of resistance: tracing transformations of power through Bedouin women. *American Ethnologist, 17*(1), pp. 41–55.

Ahmed-Ghosh, H. (2006). Voices of Afghan women: Women's rights, religion and culture. *International Feminist Journal of Politics, 8*(1), pp. 110–128.

Amiss, C. & Neale, A. (2006). Asylum from rape. *Critical Half: Bi-Annual Journal of Women for Women International, 4*(1), pp. 29–32.

Amowitz, L. Reis, C., Lyons, K. Vann, D., Mansaray, B., Akinsulure-Smith, A. et al. (2002). Prevalence of war-related sexual violence and other human rights abuses among internally displaced persons in Sierra Leone, *JAMA, 287*(4), pp. 513–527.

Bartels, S., VanRooyen, M., Leaning, J. Scott, J. & Kelly, J. (2010). *Now the world is without me: An investigation of sexual violence in eastern Democratic Republic of Congo.* Boston: Harvard Humanitarian Initiative.

Bhabha, J. (1996). Embodied rights: Gender persecution, state sovereignty, and refugees. *Public Culture, 1*(9), pp. 3–32.

Bhabha, J. (2004). Demography and rights: Women, children and access to asylum. *International Journal of Refugee Law, 16*(2), pp. 227–243.

Bhabha, J. (2007). The gendered fallout of war. Unpublished manuscript.

Bhabha, J. & Shutter, S. (1994). *Women's movement: Women under immigration, nationality and refugee law.* London: Trentham Books.

Burnett, A. & Peel, M. (2001). Health needs of asylum seekers and refugees. *British Medical Journal, 322*(7285), pp. 544–547.

Clawson, H., Duth, N., Solomon, A. & Grace, L. (2009). *Human trafficking into and within the United States: A review of the literature.* Washington, DC: Department of Health and Human Services.

Clifford, C. (2008). Rape as a weapon of war and its long-term effects on victims and society. Paper presented at 7th Global Conference: Violence and the contexts of hostility, Budapest, Hungary. Retrieved from http://www.inter-disciplinary.net/ptb/hhv/vcce/vch7/Clifford%20paper.pdf.

Courtois, C. & Ford, J. (2008). (Eds.), *Treating complex traumatic stress disorders: An evidence-based guide.* New York: Guilford.

Crawford, M. & Marecek, J. (1989). Psychology reconstructs the female. *Psychology of Women Quarterly, 13*, pp. 147–165.

Deer, S. (2010). Relocation revisited: Sex trafficking of native women in the United States. *William Mitchell Law Review, 36*, pp. 622–683.

Drozdek, B. & Wilson, J.P. (2004). Preface: Broken spirits in an era of trauma, terrorism, and globalization. In J.P. Wilson & B. Drozdek (Eds.), *Broken spirits: The treatment of traumatized asylum seekers, refugees, war and torture victims* (pp. xxvii–xxx). New York: Brunner-Routledge.

Eisenman, D., Keller, A. & Kim, G. (2000). Survivors of torture in a general medical setting: How often have patients been tortured and how often is it missed? *Western Journal of Medicine, 172*, pp. 301–303.

Farr, K. (2005). *Sex trafficking: The global market in women and children.* New York: Worth Publishing.

Ghorashi, H. (2005). Agents of change or passive victims: the impact of welfare states (the case of the Netherlands) on refugees. *Journal of Refugee Studies, 18*(2), pp. 181–198.

Godziak, E. & Bump, M. (2008). *Data and research on human trafficking: Bibliography of research-based literature.* Washington, DC: Institute for the Study of Human Migration, Georgetown University.

Hajdukowski-Ahmed, M. (2007). A dialogical approach to identity: implications for refugee women. In M. Hajdkowski-Ahmed, N. Khanlou and H. Moussa, (Eds.), *Not born a refugee woman: contesting identities, rethinking practices* (pp. 28–54). New York: Berghahn.

Herzfeld, M. (2007). Small-mindedness writ large: on the migrations and manners of prejudice. *Journal of Ethnic and Migration Studies, 33*(2), pp. 255–274.

Hooberman, J., Rosenfield, B., Lhewa, D., Rasmussen, A. & Keller, A. (2007). Classifying the torture experiences of refugees living in the Unites States. *Journal of Interpersonal Violence, 22*(1), pp. 108–123.

Hondagneu-Sotelo, P. & Avila, E. (1997). I'm here but I'm there: The meanings of Latina transnational motherhood. *Gender and Society, 11*, pp. 548–571.

Iacopino, V., Frank, M., Keller, A., Fink, S., Pallin, D., Waldman, R. et al. (1999). *War crimes in Kosovo.* Boston: Physicians for Human Rights.

Kallivayalil, D. (2013). Women seeking political asylum: Negotiating authority, gender and a lost home. *Women & Therapy, 3/4*, pp. 319–331.

Kea, P. & Robert-Holmes, G. (2013). Producing victim identities: Female genital mutiliation and the politics of asylum claims in the United Kingdom. *Identities: Global Studies in Culture and Power, 20*(1), pp. 96–113.

Kelly, J., VanRooyen, M., Kabanga, J., Maclin, B. & Mullen, C. (2011). *Hope for the future again: Tracing the effects of sexual violence and conflict on families and communities in eastern Democratic Republic of the Congo.* Boston: Harvard Humanitarian Initiative.

Kelly, N. (1993). Gender-related persecution: Assessing the asylum claims of women. *Cornell International Law Journal, 26*, pp. 625–674.

Middleton, J. (2010). Barriers to protection: Gender-related persecution and asylum in South Africa. In I. Palmary, E. Burman, K. Chantler & P. Kiguwa (Eds.), *Gender and Migration: Feminist Interventions* (pp. 67–85). London: Zed Books.

Mizus, M., Moody, M., Privado, C., Douglas, C., Collins, J., Mara E., et al. (2003). Germany, US, receive the most trafficked women. *Off our backs, 33*, pp. 7–8.

Neuwirth, J. (2008). *Statement of Jessica Neuwirth, President of Equality Now, to the New York City Council* 6/11/08.

Office of Refugee Resettlement. (2006). *Office of Refugee Resettlement (ORR) services for survivors of torture program.* Retrieved from http://www.acf.hhs.gov/programs/orr/programs/torturep.html (November 15, 2006).

Palmary, I. (2010). Sex, choice and exploitation: reflections on anti-trafficking discourse. I. Palmary, E. Burman, K. Chantler & P. Kiguwa (Eds.) *Gender and migration: Feminist interventions* (pp. 50–64). London: Zed Books.

Paris, R. (2008). "For the dream of being here, one sacrifices...": Voices of immigrant mothers in a home visiting program. *American Journal of Orthopsychiatry, 78*(2), pp. 141–151.

Peterman, A., Palermo, T. & Bredenkamp, C. (2011). Estimates and determinants of sexual violence against women in the Democratic Republic of Congo. *American Journal of Public Health, 101*(6), pp. 1060–1067.

Radan, A. (2007). Exposure to violence and expressions of resilience in Central American women survivors of war. *Journal of Aggression, Maltreatment and Trauma, 14*(3), pp. 147–164.

Reid, J. (2012). Exploratory review of route-specific, gendered and age graded dynamics of exploitation: Applying life course theory to victimization in sex trafficking in North America. *Aggression and Violent Behavior, 17*, pp. 257–271.

Schauer, E. & Wheaton, E. (2006). Sex trafficking into the United States: A literature review. *Criminal Justice Review, 31*(2), pp. 146–169.

Sircar, O. (2006). Can the women flee? Gender-based persecution, forced migration and asylum law in South Asia. In N. Behara (Ed.), *Gender, conflict and migration* (pp. 255–273). New Delhi: Sage.

Smith, H. & Keller, A. (2007). The context in which the treatment takes place: The multi-faceted stressors facing survivors of torture and refugee trauma. In *"...Like a refugee camp on 1st Avenue": Insights and Experiences from the Bellevue/NYU Program for Survivors of Torture* (1–37). New York: Author.

Stark, A. (1996). Posttraumatic stress disorder in refugee women: How to address PTSD in women who apply for political asylum under grounds of gender-specific persecution. *Georgetown Immigration Law Journal, 11*, pp. 167–198.

Thapan, M. (2006). Series introduction. In N. Behara (Ed.), *Gender, conflict and migration* (pp. 7–17). New Delhi: Sage.

Tummala-Narra. U. (2004). Mothering in a foreign land. *The American Journal of Psychoanalysis, 64*, pp. 167–182.

UNHCR (2011). Retrieved from http://www.unhcr.org/pages/49c3646c4d6.html (April 6, 2013).

Zavos, A. (2010). Gender, migration and anti-racist politics in the continued project of the nation. In I. Palmary, E. Burman, K. Chantler & P. Kiguwa (Eds.), *Gender and migration: Feminist interventions* (pp. 15–30). London: Zed Books.

Zeigler, S. & Stewart, K. (2009). Positioning women's rights within asylum policy: A feminist analysis of political persecution. *Frontiers: A Journal of Women's Studies, 30*(2), pp. 115–142.

7

Sexual Violence in an Immigrant Context: South Asian Women in the United States

Pratyusha Tummala-Narra, Anmol Satiani, and Neha Patel

Recent events, such as the brutal gang rape and subsequent death of a 23-year-old woman on a bus in New Delhi, India in December, 2012, have stirred global public outrage. Yet, violence against women, both physical and sexual, is a major public health problem worldwide, occurring across cultural and socioeconomic contexts (Basile, Chen, Black & Saltzman, 2007; Bryant-Davis, Chung & Tillman, 2009). While the problem of sexual violence is rooted in social injustice in any cultural context, it is further complicated in the case of immigrant women of color who cope with layers of inequality and discrimination within and outside of their ethnic communities. Because of these challenges, traumatic experiences of immigrant women are often unnoticed and unaddressed within families, communities and broader society. In particular, sexual violence experienced by South Asian women has been virtually unaddressed in the psychological literature, paralleling the lack of discussion and silencing of this issue in South Asian communities.

This chapter provides a framework for understanding sexual violence in the experience of South Asian women (first and later immigrant generations) in the United States from a feminist psychodynamic lens. Guided by research, clinical literature and our experiences as three South Asian women psychologists, we discuss the impact of sexual violence occurring in childhood and adulthood and the ways in which women negotiate their recovery and empowerment in the context of immigration. In the following sections, we present contextual issues contributing to the problem of sexual violence, recovery and empowerment within South Asian communities in the United States. We then present clinical vignettes that will illustrate how cultural context and social oppression shape stress and resilience in the face of sexual trauma. We end this chapter with reflections on the clinical vignettes from an integrative approach to psychotherapy with South Asian women survivors of sexual violence.

Approaching sexual violence in the South Asian context: An integrative perspective

Sexual violence against women in the South Asian community in the United States should be considered within and across multiple layers of reality. Examinations of social locations, intersections of identity and oppressions are cornerstones of contemporary feminist therapy and are critical in understanding sexual violence (Enns, 2012). Individuals' experiences of power and privilege may shift according to the context. For example, women might belong to both privileged and oppressed groups simultaneously, and these shifts impact experiences of sexual violence (Enns, 2010). Specifically, in working with South Asian trauma survivors, we know that individual experiences of sexual trauma are shaped by gender socialization, cultural and religious beliefs, experiences of acculturation and discrimination (e.g. sexism, racism, heterosexism, classism, ableism and religious bias) and access to resources. Attending to the multiple determinants of traumatic stress is critical to developing feminist, culturally responsive psychotherapy and requires a theoretical approach that emphasizes individual, interpersonal and systemic factors.

The use of masculine epistemology in descriptions of women's social and psychological experiences and neutral, decontextualized notions of the psychotherapist has been challenged by feminist psychologists (Kaschak, 1992). Feminist perspectives emphasize the role of social position and privilege in women's ability to voice the realities of traumatic experience and a personal sense of freedom within male-dominated structures that contribute to sexual violence and related isolation of women and girls (Abraham, 2006; Brown, 2010; Espín, 2006). The therapist's gender, race, religion, culture, sexual orientation, social class and dis/ability are thought to be critical to approaching power dynamics and relations in her/his relationship with the client (Kaschak, 1992; Espín, 2006). Over the past two decades, feminist, multicultural scholars (e.g. Comas-Diaz, 2006; Espín, 2006) have emphasized the recognition of women's indigenous cultural narratives that reflect both histories of colonization and oppression as well as individual and collective sources of resilience.

Scholars further note the ways in which South Asian immigrant women experience significant changes in family structure that shape experiences of femininity and of navigating multiple cultural contexts (Patel, 2007; Tummala-Narra, 2013). While we believe that feminist, multicultural perspectives are essential to understanding contextual influences on sexual violence, we integrate psychodynamic perspectives to address the intersection of sociocultural context and intrapsychic and interpersonal processes relevant to South Asian women. Psychodynamic perspectives emphasize the role of affect, defenses used to cope with distressing thoughts and feelings, identification of recurring themes and patterns, interpersonal relations and symbolic life (McWilliams, 2003; Shedler, 2010). Contemporary, relational psychodynamic perspectives focus on how traumatic and non-traumatic events shape unconscious and conscious processes that can be mirrored in the therapeutic relationship (Layton, 2004). From an integrative perspective, the subjective experiences of the client and the therapist – with an awareness of social locations, intersections of identities, power, privileges and oppressions – are thought to be critical for building a safe, collaborative relationship in which the client can cope with traumatic stress and both explore and make meaning of traumatic events. The therapeutic relationship is thought to facilitate mourning and empowerment, guided by the client's engagement with traumatic material and by creating her own personal narrative consistent with her worldview.

Taken together, feminist and contemporary psychodynamic perspectives provide a framework for addressing the complexity of sexual violence among South Asian women in the United States. Specifically, for South Asian women, a rapidly growing immigrant subgroup in the US, gender, culture, religion and class play significant roles in identity development under traumatic and non-traumatic circumstances. The term "South Asian" most typically refers to individuals whose origins were in India, Pakistan, Bangladesh, Nepal and Sri Lanka (Dasgupta, 1998; Patel, 2007). The diversity both within and across these countries of origin reflects differences in religions (e.g. Hinduism, Islam, Sikhism, Christianity, Buddhism, Jainism and Judaism), languages and customs. Further, the experience of immigration is influenced by the specific wave of migration or time period in which South Asians arrive in the United States. At the same time, South Asians share commonalities with respect to cultural values and traditions, particularly concerning gender. South Asian women contend with gendered expectations in their communities and with stereotypes imposed from mainstream, patriarchal North American society. Similar to other immigrant women, South Asian women tend to become the carriers of cultural traditions, and they bear the burden of maintaining cultural continuity and stability for their families, at times at the cost of personal freedom (Espín, 2006; Patel, 2007).

Over the past several decades, however, South Asian women have increasingly organized around the problem of domestic violence and have developed community-based strategies to provide emotional and instrumental support to women survivors of marital violence (Abraham, 2006). These developments challenge both mainstream South Asian and mainstream North American contexts, and they call attention to South Asian women's efforts to confront the silence concerning violence against women. Yet, public perceptions, including those of mental health professionals, continue to be influenced by stereotypes of South Asian women as passive, and of South Asian men as dominant and oppressive (Ahmed, Reavy & Majumdar, 2009). Feminist, multicultural scholars, in particular, have cautioned against broad attributions of male dominance to cultural values rather than to violence itself (Espín, 2010; Kallivayalil, 2008; Tummala-Narra, 2011a). These scholars call for a more in-depth analysis of the multiple factors that contribute to sexual violence against South Asian women, which recognizes the resilience of South Asian women. They further call for attention to the intersection of context and intrapsychic and interpersonal processes that reflect the experiences of South Asian women survivors of sexual trauma (Tummala-Narra, 2013).

Bearing in mind the private nature of sexual trauma in the South Asian community, we have been interested in bringing voice to South Asian women survivors and specifically how women engage with both resisting and reifying patriarchy as a way of maintaining cultural continuity in the immigration process. We recognize the dynamic nature of culture and beliefs concerning gender and violence, and as such, we emphasize both the context that shapes realities of sexual violence and the unique psychological experiences of survivors that are rooted in these realities. We take the position that epistemological approaches to sexual violence against women in the South Asian context must be rooted in the indigenous narratives of South Asian women (Kaschak, 1992; Tummala-Narra, 2011b) and that each aspect of women's psychological experience "contains all the others" (Kaschak, 1992, p. 29). As such, we emphasize the importance of intersectionality and complex interplay of intrapsychic, interpersonal and social layers of experience in the face of trauma. In the following sections, we describe in more detail the intersection of gender with cultural and religious beliefs and with immigration among South Asian women. We believe that this contextual analysis will provide the foundation for an integrative, culturally responsive approach to psychotherapy.

Gender, sexual violence and the South Asian context

Women's fear of being assaulted by men is pervasive across cultures (Herman, 1992; Kaschak, 1992). It is important to note, however, that sociocultural context shapes beliefs about gender and sexuality that influence attitudes and responses to sexual violence. In the South Asian context, religious and cultural beliefs and customs concerning gender, sexuality and the role of men and women in the family can contribute to conceptualizations of violence against women.

Gender, religion and spirituality

In South Asian cultures, there is a simultaneous idealization of women as powerful figures and the devaluation of women (Navsaria & Petersen, 2007). Indira Gandhi became prime minster of India in 1966 at a time when few Western countries would have elected a woman as head of state. Although images of idealized, powerful female figures and religious deities are common in South Asian cultures, South Asia remains a male-dominated, patriarchal culture where women are simultaneously devalued, revealing an inherent contradiction.

For example, in Hindu culture, female forms of God are honored for various qualities, such as the power to create life and to protect against evil. These female images of power are internalized by many Hindus, facilitating a sense of respect for women, especially maternal figures (Patel, 2007; Tummala-Narra, 2011a). In other instances, women and men sometimes engage in religious practices that undermine women's worth. For instance, women are considered impure when they menstruate, and as such, women are typically not allowed to attend temples or mosques during menstruation. Instead of being a source of strength and support for women, some religious practices contribute to yet more male privilege and power. Further, these contradictions pose risks for an idealization of maternal attributes, sexual objectification of women and sexual desire of women, coupled with fear of women's expressions of sexuality and power. According to Singh (2006), "The patriarchal threads of South Asian culture that accord lower status to women may weave a quilt of culturally-endorsed female responses to sexual violence" (p. 3). The silence of women's voices is one such response.

At the same time, religious and spiritual practices are experienced by many South Asian women as sources of strength and resilience, particularly in the face of adversity. Women, especially mothers, are primary transmitters of religious and spiritual practices in the realm of the family, although they are less likely than men to hold positions of authority in religious institutions (e.g. temples, mosques and churches). Engagement in religion and spirituality is often a source of pride for women, often inextricably tied to cultural identity. For example, for many Muslim women, wearing a headscarf is a source of pride and connection with religious and spiritual identity, a practice often interpreted by non-Muslims as an indication of oppression (Sirin & Fine, 2008; Tummala-Narra & Claudius, 2013).

Sexuality

South Asian cultural prescriptions generally allow men rights to premarital sex and allow sexual gratification within marriage while simultaneously defining women's sexuality in terms of premarital chastity and reproduction. Female premarital worth is traditionally defined by her sexual purity, which if violated tends to be connected with shame and sometimes social ostracism by the community (Abraham, 1999). This is related to a cultural prescription that men should control women's sexuality and women should be controlled in order to defend family honor. Additionally, there is a complex intersection of class and caste in men's accessibility to unmarried woman and girls who are seen as sexually exploitable, such as those from lower castes and income backgrounds (Abraham, 1999).

South Asian cultures tend to permit greater physical, although not sexual, affection and contact with same-sex individuals than with members of another sex in public. The only permissible context where the act of sex can be expressed between heterosexual individuals is under the sanction of marriage, and even under the sanctions of marriage there continues to be a general disapproval of the erotic aspects of marriage (Nath & Nayer, 2004). Sudir Kakar, a psychoanalyst who has studied sexuality in an Indian context, noted that sexual taboos are so strong that in some Hindu communities women's genitals do not have a name (as cited in Nath & Nayer, 2004). According to Nath and Nayer (2004), this type of repressed sexuality contributes to the problem of incest in India. Multiple factors such as greater repression of sexual impulses, silencing of women and children and abuse that's rooted in patriarchy contribute to the problems of incest, marital rape and sexual assault.

Family

South Asian cultures tend to be collectivistic in nature, emphasizing the needs of the family or group over the needs of the individual, an orientation that is even considered a South Asian "trait" (Mehrotra & Calasanti, 2010; Tummala-Narra, 2013). South Asian family structures traditionally include a patriarchal hierarchy, based on gender, age, birth order, marital status and respect for elders (Dasgupta, 1986). Further, it is common for a family member to sacrifice his or her individual wishes and desires when in conflict with the needs of the family unit, and at times, family members collectively use shame as a form of social control to maintain the structure, unity and honor of the family. Men are typically considered to be heads of the family with unchallenged authority (Ati, 2008). In contrast, although women have been increasingly working outside of the home, women typically continue to emotionally and physically care for the family. In a family structure where family unity and men's authority is valued, women's financial dependence on men and family secrecy can create a context in which violence not only takes place but is condoned (Niaz, 2003).

Although under non-traumatic circumstances, the family structure can offer stability and continuity to women and men, it contributes to challenges in acknowledging and addressing sexual violence within the family and the community. The belief that family problems should be dealt with privately, prohibitions on discussions of sex and sexuality and deference to older members of the family contribute to the silencing of sexual violence against women and girls. Women's and girls' silencing of their experience of sexual violence is a self-protective strategy and a form of resilience in a context where disclosure could be more harmful. According to Poore (2000), silence is embedded within this South Asian context and is related to patriarchal power dynamics in families (Singh, 2006). Through observation and socialization, girls learn from other women in their families to maintain silence about violence, which is meant as indicative of their strength and ability to endure hardship without complaint. Silence then becomes a strategy for empowerment and resilience for South Asian women in their families: "They learn how to engage in silent rebellion and to confront adversity with silent tactics. Silence is no longer just a survival mechanism; it becomes a testament to resilience. It is equated with inner power" (Singh, 2006, p. 9).

Sexual violence in the South Asian immigrant context

Acculturation and ethnic identity

South Asian immigrant women's experiences of violence are shaped not only by their cultural and religious traditions but also by their experiences of migration to a new country (Anitha, 2011). For many women, entering the work force in a new country contributes to important changes in family structure and dynamics, including potential shifts in power among men and women. However, it has been noted that immigrant women may view their employment as an extension of their responsibilities to their families rather than as a substitute for their domestic roles (Suh, 2007). Migration has been thought to increase "women's vulnerability to experiencing detrimental expressions of patriarchy", particularly due to economic, legal, informational, linguistic and other barriers limiting access to services and supports (Ahmad, Riaz, Barata & Stewart, 2004, p. 263). Institutional racism from service providers, lack of adequate support in terms of welfare services, public housing, health services, linguistic barriers and immigration policies are some of the structural challenges for South Asian, immigrant women in the United States (Anitha, 2011). Further, acculturation may impact conceptualizations of violence and, therefore, their disclosure and help seeking. For example, women may not view their experiences as violence, depending on their cultural beliefs including those related to patriarchy (Ahmad et al., 2004). Some research (Devdas & Rubin, 2007), in fact, suggests that perspectives on rape differ between first- and second-generation South Asian American women, where the first generation reports more rape myth acceptance and a tendency to believe that even coerced sexual activity is preventable and the responsibility of women, when compared with second-generation women.

Traumatic experiences can complicate women's perceptions of themselves as cultural beings and their identification with their ethnic groups. Sexual violence can raise conflicting feelings about women's roles and identities as South Asians. Mahalingam and Haritatos (2007), for example, observed a need for immigrants to project more positive idealized social identities as compared with those in the dominant group. Upholding such positive images can serve to counteract denigrating messages from the dominant culture. While such images of South Asian communities promote a positive self-image for South Asians, they may also obscure challenges in these communities, such as sexual violence. Women in particular are often in the position of engaging in "community impression management" (Abraham, 2006) because they are expected to be model members of both their families and communities (Kanukollu & Mahalingam, 2011; Mahalingam & Haritatos, 2007). Some South Asian women survivors of childhood sexual abuse have noted that having a strong ethnic identity allows them to develop assertiveness, pride and the ability to connect with South Asian support systems. On the other hand, experiences of discrimination diminished the energy required for healing from trauma (Singh, Hays, Chung & Watson, 2010).

Discrimination

With new norms around gender, sex, sexuality and family, South Asian women grapple with gendered racism, a term coined by Philomena Essed (1991), in which both sexist and racist ideologies are internalized from both their cultures of origin and North American society. Gendered racism is exemplified in stereotypes of South Asian American women as either hypersexualized and exotic or asexual and undesirable, as well as passive and weak. They are "othered", while white women are viewed as "normal", confident and assertive (Patel, 2007). Despite the heterogeneity of South Asian women, many North Americans perceive South Asian women collectively as described above and as docile, submissive and subservient and as a model minority (Dasgupta, 1998; Patel, 2007; Tummala-Narra, 2013). South Asian women within the context of a post-9/11 US may also experience microaggressions related to race, gender, religious, class and sexual orientation. Recent studies of discrimination among South Asians post-9/11 indicate that first- and second-generation South Asians indeed experience considerable racial and ethnic discrimination, which are connected with mental health problems such as depression (Kaduvettoor-Davidson & Inman, 2012; Tummala-Narra, Alegria & Chen, 2012). In the case of violence, racist language may be used by male perpetrators with the intention of eroding women's self-esteem and cause them to question their competence and ability to survive independently, thereby contributing to their silence (Anitha, 2011).

Larger structural problems (e.g. sexism, racism, classism, heterosexism and ableism) also pressure women to remain silent about violence. Women further contend with discrimination when seeking help from providers and agencies. Police, courts and health care providers are often unresponsive to the needs of South Asian immigrant women (Abraham, 2006). Providers also tend to have "dichotomous" and "static" understandings of culture, and they presume that the solution is for South Asian women to adopt a more "Western" lifestyle (Burr, 2002). Espín (2006) described the ways that larger systems dismiss gender oppressions in the name of cultural sensitivity. Such systemic neglect contributes to fears that sexual violence will not be taken seriously (Patel, 2007).

Social isolation and lack of access to resources

Migration itself can isolate women. In addition, perpetrators of violence can exercise intentional strategies to isolate women in a number of ways, thereby cutting them off from potential resources in the new country and in their country of origin (Anitha, 2011; Abraham, 2006). In the country of origin, their relatives and friends may have served as buffers from the violence and/or been confidantes. However, social isolation in the US can have the effect of silencing women.

For some women who have family in the US, this does not guarantee that they have support, as their families may expect them to endure violence and may not be sympathetic to their situations (Ahmed, Reavey & Majumdar, 2009). In some cases, women may be sexually abused by their spouses or other men in the family, causing shame, fear, helplessness and anger (Anitha, 2011). Women may be silent about violence due to internalized messages that men are rarely at fault, the fear that they would be blamed for the abuse, or not being believed that the abuse occurred. Some women may also fear that the violence will escalate, particularly if they have been threatened by perpetrators, feel a need to protect and remain connected to their children, have immigration concerns, financial dependence, lack of awareness of available resources and lack skills in navigating US systems (e.g. English proficiency, lack of experience with public transportation, securing bank accounts and finding employment) (Ahmad et al., 2004; Dasgupta, 2005; Gilbert, Gilbert & Sanghera, 2004; Gill, 2009). Additionally, they may have fears of cultural insensitivity by providers and especially of being offered simplistic solutions such as disconnecting from or leaving the family without understanding the complexity of their circumstances (Chew-Graham et al., 2002; Gilligan & Akhtar, 2005). The ways in which sexual violence in the country of origin has been addressed is another likely influence on women's decision to report or not report violence in the new country. For example, women and girls abused in hospitals, such as by male teachers or by police officers, can be harassed for speaking out in Pakistan, India and other places in South Asia, and so, it is conceivable that South Asian immigrants may feel a similar reluctance to report sexual violence in the US (Niaz, 2003).

Much of the attention concerning violence against South Asian women has focused on domestic violence, which may or may not include sexual violence. South Asian communities in the US have not yet collectively mobilized to address the problem of sexual violence against women and girls. In clinical practice, however, practitioners hear stories of sexual violence from South Asian women who are able to find access to and are interested in psychotherapy and counseling. In the following section, we share three vignettes to illustrate our work as psychotherapists with South Asian survivors of sexual violence, highlighting both the urgency and the complexity of this work.

Case illustrations

Case 1: Tania (Neha Patel)

Tania is an Indian American woman who's 30 years old, single and Muslim; she was born and raised in the United States and identifies more with South Asian culture than with Western European American culture. She requested a female South Asian psychotherapist, hoping that she would be better understood. When I first met Tania, she was depressed and suicidal. I suspected sexual abuse from the beginning, and when I inquired about a potential sexual trauma history, she denied it. While she was initially able to tell me about the domestic violence in her home, it was only after two years of treatment that Tania told me her father had also sexually abused her.

Tania's father had been abusive, both physically and sexually, toward Tania's mother and two sisters her whole life. Despite this, Tania deeply loved her father, explaining that the violence occurred only when he was "sick". As the eldest daughter, Tania always protected her mother and sisters during her father's manic episodes. The sexual and physical violence had been kept secret for multiple reasons: shame about the abuse, protection of family honor and fear of deportation (since her father was an undocumented immigrant). Tania also wanted an arranged marriage, and she worried that her prospects would diminish if people found out she had been sexually abused.

Trauma experts have found that healing from sexual trauma involves breaking silence about the traumatic event(s) and gaining a sense of agency. This is in direct conflict with South Asian cultural prescriptions of silence. Tania's silence had protected her from additional trauma from community stigmatization, but it also allowed the abuse to continue. As her therapist, I struggled with how to integrate traditional therapeutic approaches while being culturally sensitive.

When we first began treatment, Tania told me that I was using a "culturally insensitive" model by expecting her to show up at my office for sessions. She felt burdened by what she saw as my demand and not her need. She told me that I had to listen to what she needed, which was the flexibility to have phone sessions given her family responsibilities. At the time, I did not see this as a cultural issue in the same way that she saw it. I discussed my concerns about feeling conflicted about doing phone sessions with her given the lack of privacy in her home and the ongoing threats of possible violence. Was I colluding in the silence or adapting to her needs in a more sensitive way? I agreed to the phone sessions with the hope that Tania would feel an increased sense of control and agency in her own life and that she would feel heard and responded to. During phone sessions, she would frequently tell me how she wasn't able to talk about matters regarding her family, what was happening within her home or anything of significant emotional distress. I struggled with my own feelings of being silenced by the setup of phone sessions. Had I not offered the phone sessions, it was likely that Tania would have dropped out of therapy. I voiced these dilemmas to her, and we talked about how she felt "caged" at home and silenced in her life.

Silence and withdrawal served many functions for Tania. These strategies helped her to manage her feelings, to protect herself from further trauma, to disappear and self-destruct and/or to hurt those from whom she had pulled away. However, these strategies also left her feeling profoundly alone. She worried that even if she spoke out or screamed, no one would be able to help her. She lived in a continued state of helplessness, and the silence left her feeling imprisoned. In a sense, this paralleled my initial feelings about our arrangement for phone sessions. We talked about how that manifested in our relationship. She told me that the phone sessions allowed for her to trust that I could listen to her needs and that I could help her. As she became more connected to me, the frequency of the phone sessions decreased; she increasingly came into my office for sessions. Still, sometimes she wouldn't show up for an appointment or would cancel a session at the last minute, explaining that she had family matters to tend to.

As she felt safer in our relationship, she would often say after broaching a topic herself, "I don't want to talk about that anymore". I saw this as growth, as previously she felt little control in her interpersonal interactions and would either consciously or unconsciously avoid or change subjects to cope with her anxiety. She now felt she had the freedom to choose what she wanted to speak about and when. She began to talk more about her father's manic episodes, and it became clearer to her that he was struggling with an untreated mental illness. We discussed ways in which she could help her father get treatment since she continued to feel responsible for taking care of him and the rest of her family. I connected her with providers and resources that she felt she could trust given our relationship.

Although her father is in and out of treatment, Tania still feels relieved and less burdened in having to be present at all times to assure her family's safety, although at times she is still called to protect her family. She was able to move out of her family home and create safety for herself while still tending to her family responsibilities. She came to life as her safety and stability in our relationship grew. Tania's self-care improved; her sense of helplessness decreased; and she began to experience moments of joy in her life and experienced comfort in our relationship and in a few of her friendships. She experienced a greater sense of freedom at work without feeling trapped by her responsibilities. She gradually developed an increased sense of agency in her life. It was at this point, two years into her therapy, that Tania finally told me that she had also been sexually abused by her father. I asked her what it was like for her to tell me, and she said, "It feels like a relief" and that she didn't have to worry about the consequences of revealing that to me as our relationship was safe, unlike her relationships in her community, where she felt she would be shamed.

For Tania, it was our relationship that helped her heal and eventually to break her silence to me on her terms. Our relationship allowed her to develop a capacity for connection, safety and trust. After her disclosure, our therapy sessions continued to focus on our connection and on working through Tania's shame surrounding the sexual abuse and her fears of judgments by others. She began to feel more empowered in creating safe and fulfilling relationships with others, and her profound loneliness and depression had gradually dissipated with a new community of relationships she had begun to build for herself.

Case 2: Reena (Anmol Satiani)

In a community mental health center, I worked with Reena, an Indian immigrant woman who is in her 40s, poor, non-English speaking and Hindu. She was a mother of several children, originally from a small village in India, now living in a large, metropolitan US city for the last few years. When I met her, Reena was smiling warmly, despite suffering from a life-threatening illness and being grossly underweight as a result. She was introduced to me by a case worker with a child protective agency, who was involved with the family because one of Reena's daughters had reported that her father had sexually abused her. When we discussed her concerns, Reena focused on not having food, not being able to pay bills, not having a phone, being unemployed and not having sufficient medical care. Related to these worries, she expressed linguistic barriers, transportation concerns and a lack of knowledge of available resources. After several months of working to connect her and her children to needed services, our relationship deepened. Reena disclosed that she was being raped by her husband almost daily and disclosed her worries that he also may have sexually abused their older daughters. Although she expressed worry about her children and presented what would be described as symptoms of depression and post-traumatic stress disorder (PTSD), her daily experiences of violence and symptomology were not her "presenting concerns".

When I asked about her understanding of the violence, Reena told me that she was being inhabited by a jinn or spirit. Her solution to this problem was to go to India and have the spirit removed by a *pundit* or Hindu priest. We did discuss our differing understandings of violence at the time and I was able to share that I did not conceptualize her suffering in the same manner. Given how important this was to her, particularly in terms of her religious identity, I decided to support her in going to India to have this jinn removed from her body, but also worked to help to keep her engaged in therapy. She did eventually go to the *pundit* in India, but she later returned and commented that this was not effective given that the violence persisted after she returned. I believed that this represented a shift in her location of the problem as lying inside her to the possibility of other conceptualizations of the violence.

I was concerned with the ways that both her family and cultural communities and US systems reinscribed her experience of oppression, rendering her invisible and without voice and resources. Her hopelessness was profound and seemed to render her powerless and distanced from her internal voice or instincts. We worked to consider ways that she could gain a sense of personal power and gain some control in her life, consistent with feminist empowerment approaches (Worrell & Remer, 2003). One example of this was that I initially did home visits, and when this became too dangerous for me (Reena's husband threatened to harm me and my colleagues), I encouraged her to come to the agency via public transportation to meet with me. Interestingly, Reena expressed feeling positive about this change as she became much more comfortable and confident in her ability to navigate the city. At times, she entered the agency, dressed formally in Indian clothing and holding hands with her children, who would play in our waiting area. A couple of her children explained to their respective therapists that the agency became a safe place where they felt supported.

During one session after almost one year of working together and after her trust in me had grown, Reena and I had been discussing a safety plan for her and her children as we did frequently depending on the circumstances at the time. She asked me in Hindi, "Anmol, why do you keep asking me about the rape? Every woman in my village is raped by her husband." At this point, although I was deeply saddened and overwhelmed by this statement, I listened to her share her experiences in her village in an effort to learn and give authority to her lived experiences as told in her own voice and in her own language (Brabeck & Brown, 1997). She discussed the norms related to violence, gender roles in marriages and her gender socialization. Through this discussion, I realized that she had lived in a state of terror for many years prior to migrating to the US and that this had profound implications for her current situation. We discussed her challenges as an immigrant given her lack of supports in the US, her limited access to resources and her immigration concerns, and we discussed how these challenges related to her fears about leaving her husband. She explored related internal dilemmas – for example, worries about separating her family members from one another and how this conflicted with her ideas of what it means to be a wife and mother. This was the first time that she raised the idea of leaving her husband. She was able to discuss her worries about her children, particularly her daughters. I also acknowledged and discussed with her my positions of power and privilege – particularly as a second-generation, middle class, educated, English speaking woman – and some of my challenges in being able to fully understand her experiences. I explained my dilemma in that I did want to continue to understand her perspectives on violence but that I also disagreed with the perspective that the problem was located within her and that it should be accepted; I suggested, rather, that it was rooted in a larger system of oppression against women that needed to be addressed. I stated that I was not able to use "culture" as a justification for violence. I expressed hope for her and for her children to live in a violence-free household, and I expressed that I thought that this was possible. She smiled and shook her head as if to say, "you don't understand," although my sense was that she was pleased to hear this message from me. Through this conversation, I hoped to convey that I did not think that she was internally "bad" but that she was a woman with a great deal of strength who was capable of achieving her goals. At this point in our work, I shared that I believed that we had learned more about the strengths intrinsic to her or her "mother tongue" (Brown, 1994). This "mother tongue" encompassed her emotional strengths and resilience, her devotion to her children, her determination to thrive in her new country, her wise sense of humor, her ability to form strong and satisfying relationships with others, her concrete skills of adaptation and simultaneous maintenance of a strong connection with her roots and her religious faith.

Exploration and discussion of religion and spirituality were critical in our work and seemed essential in developing trust. Reena was deeply connected to her spiritual and religious beliefs and raised these issues each time we met. Although we shared the same religious background, our ways of understanding our religion differed. For example, her notions of *karma* were different from my own and seemed to be very much tied to her experiences of powerlessness and oppression. She focused on the idea that she needed to wait until the next life for her circumstances to improve because there was nothing to be done in this life. Reena's perspective on this issue seemed to shift with time as she felt more hopeful and capable and as she gained access to important resources. She used her spirituality and religion as resources and developed more faith that she and her children could have a promising future. Reena was living with her daughters, away from her husband, at the end of our two-year relationship.

Case 3: Ann (Pratyusha Tummala-Narra)

I worked with Ann, a 23-year-old woman, for one and half years in weekly individual psychotherapy. Her parents immigrated to the United States from Sri Lanka in the 1980s after being sponsored by family members. She and her two older brothers were born and raised in the United States. Ann had not worked with a therapist previously, but she was eager to do so, particularly one of South Asian origin, to help cope with her anxious and depressed mood,and difficulty with sleep. The intensity of her symptoms had been increasing for several months, and the typical ways in which she coped with stress (e.g. exercise, prayer and spending time with her family) did not seem to help her any longer.

When I met Ann, she described her family as supportive and felt especially close to both of her parents whom she described as somewhat "overprotective". At times, she would compare her average performance in school and in college with the achievements of her brothers whom she described as "brilliant". Ann's friendships with other girls while growing up derived primarily from her church, which she attended regularly. She described herself as a "devout Catholic", deeply influenced by her parents who also grew up as Catholic Christians in Sri Lanka. Although she identifies as Catholic, Ann stated that she feels less connected with her Sri Lankan cultural heritage. Her parents, while sponsored by family members to obtain legal residence in the United States, eventually moved away from these relatives to find employment in a different town where they had little access to interacting with other Sri Lankans. When she first entered college, Ann experienced herself as different from other South Asians and found it challenging to cope with stereotypes of South Asian women as passive and exotic. She viewed herself as an average student, contrary to the notion of a model minority, which is often applied to and internalized by Asian Americans. She also expressed her frustration with others' assuming that she is Indian and not seeing her as a Sri Lankan. Ann tended to maintain her distance from other South Asians on campus and developed friendships primarily with white European American women. She had not dated prior to college and had realized that she is attracted to women. However, she chose not to disclose this to anyone in her family or to her friends. Several years later, in her senior year, Ann attended a party where she met a white European American man whom she had seen in one of her classes. Later that evening, this man raped Ann, leaving her with significant bruising. With a friend's help, she sought medical attention but decided not to tell anyone in her family about the rape. She withdrew from most social interactions outside of her classes, and she decided not to pursue graduate education as she had previously planned.

The early part of our work together in psychotherapy focused on Ann's ability to reach out to friends and access support from people whom she experienced as safe and supportive and on her difficulty with sleep. As she came to experience her sleep as more stable, she talked in more depth about her experience of being raped and her fear about disclosing what had happened to her parents and brothers. She stated, "They wouldn't have approved of me being at this kind of party. There was a lot of drinking, but I wasn't drinking much at all. They would think that this is my fault. It would be too painful for them too. I don't think they can handle it." Ann struggled with her parents' views concerning premarital sex and homosexuality; they disapproved of both. Premarital sex was considered "bad" and issues of sexual violence were not discussed in her family or in her church. When I asked her to tell me more about the ways in which she viewed the rape, Ann stated, "I don't know. I know it's a violent act. I hate the guy who did this. At least I don't have to see him, but I just feel like the whole thing just changed me. I'm ashamed for me, what this could mean for my family if I tell them." Ann further stated that there was an unspoken view within her family and the church that homosexuality is a sin.

In the following session, Ann reluctantly stated that she had a dream that I had told her that she should tell her family about the rape, after which she felt angry. When I asked her why she had been worried about telling me about the dream, she expressed her worry about disappointing me. She stated, "It's sometimes hard for me to talk about the rape, I mean with an Indian woman." Ann further indicated that she sometimes worried that I may judge her negatively for having been at the party where she had been raped and for being attracted to women. However, at other times, she experienced me as a safe, maternal figure who provided a space for her to be herself. My own reactions to Ann's rape included feelings of sadness, anger and helplessness. I identified with her sense of loneliness as she coped with her traumatic experience without the help of loved ones and the challenge of disclosing to South Asians her sexual attraction to women. As a child immigrant to the United States from India, I was familiar with the use of silence about sexual identity and experience with sexual violence as a way to maintain connection with family and community. I also hoped to find ways to respect her conflict with difference, particularly related to her feelings about working with an Indian woman. Ann had told me about her feelings of ambivalence toward South Asians in the US, and at times she struggled with teasing me apart from South Asians, in particular Indians, to whom her Sri Lankan heritage was invisible. At several points in our work together, I told Ann that it was indeed not typical for two South Asian women to talk openly about sexuality and sexual violence and that I imagined that the process of talking about these issues would be a difficult one. At one such moment, Ann stated, "It's hard to break the silence."

As psychotherapy progressed, we continued to talk about her views of men and women as shaped by her Sri Lankan and Christian identities. Although she identified primarily as a Christian, her traumatic experience and her growing awareness of her sexual identity had raised questions about her connections with Sri Lankan culture, which she had dismissed for most of her life. Specifically, she came to understand that her fears about disclosing the sexual violence and her sexual attraction to women to her family was rooted in multiple layers of experience as a Catholic, Sri Lankan American woman. Gradually, she decided to tell one of her brothers about the rape to "test out" what it may be like to tell her family, and to her relief, she found him to be supportive. Later, she told one of her closest friends that she is attracted to women, after which the friend comforted Ann by telling her that she had "suspected this". Reconnecting with loved ones became critical to her recovery and to a related exploration of her sexuality and her cultural identity, which are both distinct from and intertwined with the rape.

Reflections on psychotherapy

The case examples (Tania, Reena and Ann) illustrate the unique experiences of South Asian women survivors of sexual violence and approaches to psychotherapy that consider the interplay of systemic (including cultural), interpersonal and individual factors in trauma recovery. All three clients struggle with the aftermath of traumatic stress that overlaps with survivors of sexual violence across contexts, as evidenced in the patriarchal assertion of power and privilege and the survivor's related traumatic stress (e.g. depressed mood, anxiety and difficulty with trusting others). However, the ways in which they cope with and make meaning of traumatic experience is embedded in their family and cultural contexts, and these vary considerably across immigrant generations, religious beliefs, social classs, sexual identities and immigration statuses. These cases underscore the diversity of experiences among South Asian women in the United States and the importance of attending to the multiplicity of identity as it bears a mutual influence with traumatic stress (Tummala-Narra, 2011b). Relatedly, the therapist's cultural context has a significant presence in the therapeutic dyad. Contrary to traditional notions of the therapist as a neutral figure, each of our case examples emphasizes the importance of the therapist's examination of her own bicultural conflicts and dilemmas and her conscious choice to listen to and privilege the client's narrative and choices. The cases further illustrate the importance of listening and bearing witness to stories of violence and terror supported within a patriarchal structure and the need for therapists and clients to engage in an analysis of power and gender. This analysis is critical given the ways in which violence is gendered and in order to highlight the powerful myths in society that contribute to victim blaming and to self-blame (Brown, 1994; Worrell & Remer, 2003). The therapeutic work required that each of us as therapists challenge our own beliefs and hold the complexity of our clients' internal and external realities. While each of our clients worked toward healing and empowerment, we have been transformed by this work.

An integrative psychotherapeutic approach that espouses feminist and contemporary psychodynamic perspectives was especially helpful in creating a therapeutic space in which the therapist and the client could work collaboratively to examine the influence of broader contextual factors on their traumatic experiences (Bryant-Davis et al., 2009; Kallivayalil, 2008). Further, we note the ways in which cultural dynamics rooted in sociohistorical differences between Hindus and Muslims and between Indians and Sri Lankans may be mirrored in the therapeutic interaction. This is also evident in differences across social class and sexual identity between the therapist and the client. An integrative approach creates opportunity for dialogue on sexuality and violence that typically does not occur in South Asian families and communities, therefore breaking the silence. This approach requires attention to dynamics of power and privilege; cultural and religious differences and similarities; and transference and counter-transference that is based in traumatic and non-traumatic life experiences. For example, each of the clients discussed the role of racism and stereotyping in their lives and sought out therapists of a familiar cultural background with whom they felt safe. However, each client still struggled with concerns about whether or not the therapist would understand her, such as in the case of Ann, who was unsure whether an Indian American heterosexual therapist would be able to truly understand the experience of a Catholic, Sri Lankan American woman who was raped and is sexually attracted to women. The client's ambivalence about deepening her relationship with her therapist was not only rooted in her difficulty with trusting others due to her traumatic experiences but also in conscious and unconscious fears that the therapist may in fact reenact the client's past and ongoing experiences of disconnection and rejection by others who don't fully recognize the complexity of her internal experience and her external realities.

A critical aspect of recovery from trauma concerns growing recognition of one's own needs, which under traumatic circumstances, would have compromised physical and emotional survival (Herman, 1997). The case examples illustrate that empowerment is inextricably tied with cultural and religious values and with the realities of cultural transition and adjustment to a new country. While the therapist and the client may differ in their views of what constitutes trauma recovery and empowerment, psychotherapy should facilitate reflection on the complexity of gender, trauma, discrimination, social position and privilege, cultural adjustment, cultural and religious identity and sexuality as well as the ability to bear this complexity and its accompanying feelings of ambivalence. It is important that the therapist recognize the dilemmas that South Asian women face with respect to prioritizing their personal needs over those of their families and communities because the latter are often deeply connected with a cohesive sense of self and cultural continuity. Thus, feminist notions of recovery and empowerment may need to be tailored to all individuals within their respective contexts. One individual's notion of moving toward recovery or gaining some control may involve, for example, taking public transportation on her own, whereas for another, it might mean telling a brother about one's experience of rape, though both may be related to the dilemmas described above.

We recognize that the experience of sexual violence in the South Asian community in the United States should receive far greater attention from clinicians and researchers. The cases presented in the chapter demonstrate a movement away from externally defined images of South Asian women and toward self-definition and self-scripting of experience (Collins, 2009). We are also aware of our own unique position as South Asian women therapists, and as such, we shared briefly some of our experiences of working with survivors who found access to psychotherapy, in the hope of breaking silence on the problem of sexual violence and raising awareness of the complexity of stress and resilience among South Asian women survivors.

References

Abraham, M. (1999). Sexual abuse in South Asian immigrant marriages. *Violence Against Women, 5*(6), pp. 591–618.

Abraham, M. (2006). Model minority and marital violence: South Asian immigrants in the United States. In R. Mahalingam (Ed.), *Cultural psychology of immigrants* (pp. 197–216). Mahwah, NJ: Lawrence Erlbaum Associates.

Ahmad, F., Riaz, S., Barata, P. & Stewart, D.E. (2004). Patriarchal beliefs and perceptions of abuse among South Asian immigrant women. *Violence Against Women, 10*, pp. 262– 282.

Ahmed, B., Reavey, P. & Majumdar, A. (2009). Constructions of 'culture' in accounts of South Asian women survivors of sexual violence. *Feminism & Psychology, 19*, pp. 7–28.

Anitha, S. (2011). Legislating gender inequalities: The nature and patterns of domestic violence experienced by South Asian women with insecure immigration status in the United Kingdom. *Violence Against Women, 17*, pp. 1260–1285.

Ati, R. (2008). *Intra Family Sexual Abuse (IFSA) in the South Asian Indian Family System: "A comprehensive Survey of the Literature and Critical Analysis."* Doctoral Dissertation, American School of Professional Psychology, Argosy University.

Basile, K.C., Chen, J., Black, M.C. & Saltzman, L.E. (2007). Prevalence and characteristics of sexual violence victimization among U.S. adults, 2001–2003. *Violence and Victims, 22*(4), pp. 437–448.

Brabeck, M. & Brown, L.S. (1997). Feminist theory and psychological practice. In J. Worell & N. Johnson (Eds.), *Shaping the future of feminist psychology: Education, research and practice* (pp. 15–36). Washington, DC: American Psychological Association.

Brown, L.S. (1994). *Subversive dialogues*. New York: BasicBooks.

Brown, L.S. (2010). *Feminist therapy*. Washington, DC: American Psychological Association.

Bryant-Davis, T., Chung, H. & Tillman, S. (2009). From the margins to the center: Ethnic minority women and the mental health effects of sexual assault. *Trauma, Violence & Abuse, 10*(4), pp. 330–357.

Burr, J. (2002). Cultural stereotypes of women from South Asian communities: Mental health care professionals' explanations for patterns of suicide and depression. *Social Science & Medicine, 55*(5), pp. 835–845.

Chew-Graham, C., Bashir, C., Chantler, K., Burman, E. & Batsleer, J. (2002). South Asian women, psychological distress and self-harm: Lessons for primary care trusts. *Health and Social Care in the Community, 10*(5), pp. 339–347.

Collins, P.H. (2009). *Black feminist thought: Knowledge, consciousness, and the politics of empowerment*. New York: Routledge.

Comas-Diaz, L. (2006). Latino healing: The integration of ethnic psychology into psychotherapy. *Psychotherapy: Theory, Research, Practice, Training, 43*(4), pp. 436–453.

Dasgupta, S.D. (1986). Marching to a different drummer?: Sex roles of Asian Indian Women in the United States. *Women & Therapy, 5*, pp. 297–311.

Dasgupta, S.D. (1998). Gender roles and cultural continuity in the Asian Indian immigrant community in the U.S. *Sex Roles, 38*(11/12), pp. 953–974.

Dasgupta, S.D. (2005). Women's realities: Defining violence against women by immigration, race, and class. In N.J. Sokoloff & C. Pratt (Eds.), *Domestic violence at the margines: Readings on race, class, gender, and culture* (pp. 56–70). New Brunswick, NJ: Rutgers University Press.

Devdas, N.R. & Rubin, L.J. (2007). Rape myth acceptance among first- and second-generation South Asian women. *Sex Roles, 56*, pp. 701–705.

Enns, C.Z. (2010). Locational feminisms and feminist social identity analysis. *Professional Psychology: Research and Practice, 41*, pp. 333–339.

Enns, C.Z. (2012). Feminist approaches to counseling. In E.M. Altmaier & J.C. Hansen (Eds.), *The Oxford handbook of counseling psychology* (pp. 434–459). Oxford: Oxford University Press.

Espín, O.M. (2006). Gender, sexuality, language, and migration. In R. Mahalingam (Ed.), *Cultural psychology of immigrants* (pp. 241–258). Mahwah, NJ: Lawrence Erlbaum Associates.

Espín, O.M. (2010, June). *Is Domestic violence a cultural tradition?...and other questions about gender and migration.* U.S. Embassy and Women without Borders, Vienna, Austria.

Essed, P. (1991). *Understanding everyday racism*. Newbury Park, CA: Sage.

Gilbert, P., Gilbert, J. & Sanghera, J. (2004). A Focus group exploration of the impact of izzat, shame, subordination and entrapment on mental health and service use in South Asian women living in Derby. *Mental Health, Religion & Culture, 7*(2), pp. 109–130.

Gill, A. (2009). Narratives of survival: South Asian women's experiences of rape. In M.A.H. Horvath & J.M. Brown (Eds.), *Rape: Challenging contemporary thinking* (pp. 161–183). Portland, OR: Willan Publishing.

Gilligan, P. & Akhtar, S. (2005). Child sexual abuse among Asian communities: Developing materials to raise awareness in Bradford. *Practice, 17*, pp. 264–284.

Herman, J.L. (1997). *Trauma and recovery*. New York: Basic Books.

Kaduvettoor-Davidson, A. & Inman, A.G. (2012). South Asian Americans: Perceived discrimination, stress, and well-being. *Asian American Journal of Psychology,* Advance online publication. doi:10.1037/a0030634.

Kallivayalil, D. (2008). Feminist therapy: Its use and implications for South Asian immigrant survivors of domestic violence. *Women & Therapy, 30*(3/4), pp. 109–127.

Kanukollu, S.N. & Mahalingam, R. (2011). The Idealized cultural identities model on help- seeking and child sexual abuse: A Conceptual model for contextualizing perceptions and experiences of South Asian Americans. *Journal of Child Sexual Abuse, 20*(2), pp. 218–243.

Kaschak, E. (1992). *Engendered lives: A new psychology of women's experience*. New York: Basic Books.

Layton, L. (2004). Dreams of America/American dreams. *Psychoanalytic Dialogues, 14*(2), pp. 233– 254.

Mahalingam, R. & Haritatos, J. (2007). Cultural psychology of gender and immigration. In R. Mahalingam (Ed.), *Cultural psychology of immigrants* (pp. 259–273). Mahwah, NJ: Lawrence Erlbaum Associates.

McWilliams, N. (2003). The educative aspects of psychoanalysis. *Psychoanalytic Psychology, 20*(2), pp. 245–260.

Mehrotra, M. & Calasanti, T.M. (2010). The family as a site for gendered ethnic identity work among Asian Indian immigrants. *Journal of Family Issues, 31*(6), pp. 778–807.

Nath, J.K. & Nayer, V.R. (2004). India. In R.T. Francoeur & R.J. Noonan (Eds.), *The Continuum international encyclopedia of sexuality* (pp. 516–532). New York: The Continuum International Publishing Group.

Navsaria, N. & Petersen, S. (2007). Finding a voice in Shakti: A therapeutic approach for Hindu Indian women. *Women & Therapy, 30*(3/4), pp. 161–175.

Niaz, U. (2003). Violence against women in South Asian countries. *Archives of Women's Mental Health, 6*, pp. 173–184. doi:10.1007/s00737-003-0171-9.

Patel, N.R. (2007). The construction of South Asian-American womanhood. *Women & Therapy, 30*(3/4), pp. 51–61.

Poore, G. (2000). *The children we sacrifice: A resource book*. Silver Spring, MD: Shakti Productions.

Shedler, J. (2010). The efficacy of psychodynamic psychotherapy. *American Psychologist, 65*(2), pp. 98–109.

Singh, A.A. (2006). *Resilience strategies of South Asian women who have survived child sexual abuse*. Doctoral Dissertation, Georgia State University.

Singh, A.A., Hays, D.G., Chung, Y.B. & Watson, L. (2010). South Asian immigrant women who have survived child sexual abuse: Resilience and healing. *Violence Against Women, 16*, pp. 444–458. doi:10.1177/1077801210363976.

Sirin, S. & Fine, M. (2008). *Muslim American youth: Understanding hyphenated identities through multiple methods*. NY: New York University Press.

Suh, S.H. (2007). Too maternal and not womanly enough: Asian-American women's gender identity conflict. In D.M. Kawahara & O.M. Espín (Eds.), *Feminist reflections on growth and transformation: Asian American women in therapy* (pp. 35–50). Binghamton, NY: The Haworth Press.

Tummala-Narra, P. (2011a). A psychodynamic perspective on the negotiation of prejudice among immigrant women. *Women &Therapy, 34*(4), pp. 429–446.

Tummala-Narra, P. (2011b). A psychodynamic approach to recovery from sexual assault. In T. Bryant-Davis (Ed.), *Surviving Sexual Violence: A Guide to Recovery and Empowerment* (pp. 236–255). Lanham, MD: Rowman & Littlefield Publishers, Inc.

Tummala-Narra, P., Alegria, M., Chen, C. (2012). Perceived discrimination, acculturative stress, and depression among South Asians: Mixed findings. *Asian American Journal of Psychology, Special Issue: Secondary analysis of the National Latino Asian American Study (NLAAS) Dataset-Part I, 3*(1), pp. 3–16.

Tummala-Narra, P. & Claudius, M. (2013). A qualitative examination of Muslim graduate international students' experiences in the United States. *International Perspectives in Psychology: Research, Practice, Consultation, 2*(2), pp. 132–147.

Tummala-Narra, P. (2013). Psychotherapy with South Asian women: Dilemmas of the immigrant and first generations. *Women & Therapy, 36*(3–4), pp. 176–197.

Worrell, J. & Remer, P. (2003). *Feminist perspectives in therapy: Empowering diverse women* (2nd ed.). Hoboken, NJ: John Wiley & Sons.

8
Supporting Transgender Immigrant Latinas: The Case of Erika

Danielle Quintero, Alison Cerezo, Alejandro Morales and Stephanie Rothman

It is estimated that the number of Latina/o immigrants who identify as lesbian, gay, bisexual, transgender or queer (LGBTQ) is nearly one million, with a quarter of them living as undocumented in the US (Center for American Progress, 2013). However, no statistic currently exists on the exact number of LGBTQ Latina/o immigrants residing in the US, which may be related to a lack of sexual orientation and gender identity questions on the US. Census as well as the fear many LGBTQ individuals face with "outing" themselves in a formal manner. While representation and consideration of LGBTQ Latina/o needs remains a serious issue, there is even less known about transgender Latina/os, especially those who are migrants to the US. The severe paucity of scholarship and theory in the mental health field on this community is especially troublesome when you consider the severe marginalization many members of this community face, including death for their gender expression (Anderson, 2010; Chávez 2010; Seuffert, 2009) and the types and amount of mental health resources needed to provide adequate emotional support and advocacy. As a direct response to this lack of focus in the mental health field, this chapter provides an ecologically based case study analysis of one woman's life experiences as a transgender immigrant Latina, which involves her migration to the US from Honduras as a means to access resources and community support related to her transgender identity.

The United States is comprised of immigrants from across the globe. According to the U.S. Census Bureau (2011), there are approximately 39.9 million immigrants who reside in the US, which includes a substantial portion of individuals who migrate from Latin America. Recent data show that Latina/o immigrants are the fastest growing ethnic group in the US and consist of 16.7% of the total US population (Pew Hispanic Center, 2012). Many immigrants arrive in the US from countries and regions that include Mexico and both Central and South America. Therefore, the diverse backgrounds of the Latina/o immigrant population present a wide range of cultural, linguistic, economic and sociopolitical backgrounds and needs.

Factors that influence motivation to migrate to the US include economic, familial and political reasons (Santiago-Rivera, Arredondo & Gallardo-Cooper, 2002; Suarez-Orozco & Suarez-Orozco, 2001; Suarez-Orozco, Suarez-Orozco & Tordova, 2008; Yakushko, Watson & Thompson, 2008; Yakushko, Backhaus, Watson, Ngaruiya & Gonzales, 2008). Although research in the mental health field that focuses on Latina/o immigrants has grown significantly in recent years, limited research focuses on the experiences and needs of LGBTQ Latina/o immigrants (Morales, Corbin-Gutierrez & Wang, 2013). The available literature on Latina/o LGBTQ people shows that they frequently migrate to the US as young adults for reasons that include access to job opportunities, healthcare and the freedom to express themselves as sexual and gender-diverse individuals (Cantú, 2009; Morales, Corbin-Gutierrez & Wang, 2013). Freedom to express one's LGBTQ identity is often directly linked to access to job opportunities, healthcare and social support in one's native countries; many LGBTQ immigrants faced severe marginalization in their native countries for being LGBTQ, which prompted their desire, and often need, to migrate to the US (Morales, 2013).

The significant lack of empirical research on transgender immigrant Latina/os is an especially serious issue as clinicians, scholars and students lack the necessary knowledge and skills to address the needs of this marginalized community. In response to this need in the field of psychology, the purpose of this chapter is to apply ecological psychotherapy (Conyne & Cook, 2004), based on Bronfenbrenner's ecological model (1979), as a guiding approach for working with Erika, a transgender (Male to Female [MtF]) woman who immigrated to the US from Honduras. In Erika's case, as well as in many other cases of transgender individuals, considering various and sometimes overlapping environmental systems is critical because they must negotiate many social networks, public institutions, policies and cultural norms in relation to the multiple facets of their identities, many of which (race, ethnicity, gender and immigration) face historical marginalization in the US. An ecological approach allows for the full consideration of the factors impacting transgender individuals' lives, thus providing a context to understand and intervene in ways that are consistent with a social justice perspective (Cerezo, O'Neil & McWhirter, 2009; Yakushko & Chronister, 2005).

In this chapter we specifically address the lived experiences of one transgender immigrant Latina, Erika (pseudonym). Erika was part of a larger qualitative project that was conducted in collaboration with a community-based organization that predominately serves transgender women who migrate to the US from Latin America. Given that this chapter is based on one woman's life, it is important to point out that no generalizations can be made from Erika's lived experiences to other immigrants, Latinas and/or transgender individuals. A sole case study was explored in depth as a means to provide a useful, comprehensive analysis of the factors impacting Erika's life, which were myriad and overtimes overlapping. Furthermore, the sheer lack of empirical research on transgender people, especially immigrants, called for a more focused exploration of one woman's life story.

Linking transgender identity and immigration

Transgender is an umbrella term used to describe "persons whose *gender identity, gender expression* or behavior does not conform to that typically associated with the sex to which they were assigned at birth", whereas gender identity "refers to a person's internal sense of being male, female, or something else" (American Psychological Association [APA], 2011, p. 1). A significant experience in the lives of many transgender individuals is the process of *transition*, wherein the individual engages in an emotional and sometimes physical transition to match their gender identity. Transition may or may not include name and gender pronoun change, alteration in physical appearance and/or medical interventions such as hormone treatments and surgical procedures. Also, it is important to point out that not all transgender people transition in a uniform manner, nor do all transgender individuals seek medical intervention (e.g. sexual reassignment surgery). Rather, when working with transgender clients, it is critical that mental health professionals use specific language (e.g. preferred name and gender pronoun) and connect with clients in a manner that is consistent with the clients' stated gender identity.

Many transgender individuals face psychological challenges on a regular basis, often solely based on their gender identity, which include stigma, bias, discrimination and a lack of social support (APA, 2011). In 2012, US Vice President Joe Biden stated that transgender discrimination is "the civil rights issue of our time" (Huffington Post, 2012), thus highlighting the increased recognition of transgender people as well as their continued experience of bias and discrimination as a common experience of life. Transphobia is defined as

> the irrational fear and hatred of all those individuals who transgress, violate, or blur the dominant gender categories in a given society, which may be experienced by transgender individuals in different ways from microaggressions to violence. (Association for Lesbian, Gay, Bisexual, Transgender Issues in Counseling, 2009, p. 4)

It is important to keep in mind that acceptance and/or discriminatory experiences transgender individuals face often differ significantly by location. Transgender Europe (TGEU), an international non-profit organization, found in its first worldwide transgender murder monitoring project that a transgender person is killed every three days in Latin America because of their gender nonconformity, which is likely to be a precipitating factor in the choice to migrate, sometimes without legal documentation (TGEU, 2009).

Transgender immigrant Latinas face particularly dangerous conditions because they are frequent targets of transphobia and xenophobia that may heavily affect their overall well-being and mental health. Similarly to the concept of transphobia, xenophobia is described as an individual having fear of or expressing hostility toward outsiders (Reynolds, Falger & Vine, 1987) or a dislike for foreign-born individuals that is based on fear or hate (Crowther, 1995). Transgender immigrant Latinas face especially severe violence related to their gender identity, including maltreatment in the forms of sexual, physical and emotional abuse during migration to the US and other countries (Anderson, 2010; Chávez, 2010). The death of Victoria Arellano in 2007, a 23-year old HIV positive undocumented transgender woman from Mexico at an Immigration and Customs Enforcement (ICE) detention center in Southern California, drew national headlines from human rights groups. Ms. Arellano was denied prescription medication to prevent the development of pneumonia that killed her two months following detainment, which many hypothesized to be related to transphobic discrimination (Chávez, 2010). Another common experience many transgender immigrant women report in detention centers is sexual assault, after which known assailants, including other detainees and sometimes even prison guards, receive no legal repercussions for their actions (Anderson, 2010; Chávez, 2010).

Once in the US, a major challenge for many transgender immigrants is securing documentation – legal permission to reside and/or work in one's adopted country. Documentation is a fundamental necessity in many counties (including the US) because it is directly tied to one's ability to earn a living and secure housing, among other important needs (Seuffert,

2009). Many transgender immigrants face the dilemma of migrating out of necessity to escape discriminatory treatment from their country, community and loved ones (Seuffert, 2009). Thus, many transgender immigrants migrate to the US with few resources for following formal legal procedures and/or lack support from family members who could help to ensure a safer migratory experience. Another challenge is that, for some transgender immigrants, the documentation they do have does not match their gender identity, thus placing them in the challenging situation of having to 'out' themselves when seeking employment. This is a serious concern as evidenced by research that supports the high degree of discrimination transgender people face, such as frequently not being hired for a job (Sangganjanavanich & Cavazos, 2010) – or in the case of immigrants – losing a job due to citizenship status (National Center for Transgender Equality, 2011).

Outside of the realm of work, transgender immigrant Latinas in the US face other serious obstacles that include susceptibility to sexual assault, police brutality and sexually transmitted infections (Center for Disease Control, 2008), and they are less likely to seek mental health and medical services as a result of their fear of deportation (Ritter & Terndrup, 2002). Although psychotherapy with transgender immigrant Latinas is becoming more relevant to the field of psychology as the cultural tapestry of the US shifts in dramatic fashion, there remains a significant dearth of empirical research that explores the lived experiences and consequent mental health needs of these women and that can help mental health professionals better understand and meet the myriad needs of this underserved community. Thus, the purpose of this chapter is to help illuminate the needs of transgender immigrant Latinas by recounting Erika's story. We intentionally used an ecological framework (Bronfenbrenner, 1979) to point out the various environmental factors impacting Erika's life, from interpersonal aspects (race and ethnicity, gender identity and sexual orientation) to societal messages related to transgender and immigrant people, as a manner to uplift the voice of one transgender immigrant Latina.

Ecological psychotherapy with transgender immigrant Latinas

Bronfenbrenner's (1979) ecological model is a useful theoretical tool for examining the complex intersections of sociocultural contexts in client's lives and has proven useful for understanding clients from historically marginalized groups (Cerezo, O'Neil & McWhirter, 2009; Yakushko & Chronister, 2005). In the case of transgender immigrant Latinas living in the US, the model helps to explain how women's life goals and behaviors, such as occupational choice and work history, are shaped by various ecological spheres and interactions between these spheres (e.g. family, neighborhood community, public institutions and popular media). For example, how messages from popular media, which often portray transgender women in a negative light (Gainor, 2000), compete or align with beliefs about transgender people from women's family of origin and peer groups. Thus, factors present within multiple ecological spheres, such as cultural messages about transgender people and/or laws and policies related to gender identity, serve as a 'cultural backdrop' that directly influences society's views on transgender (and immigrant) identity, thereby impacting women's ability to transition in a manner that is safe and accepted by the important relationships and institutions in their lives.

The ecological model is comprised of five interrelated and bidirectional systems. The *individual* system is the core of the model and consists of intrapersonal characteristics of an individual. In the case of transgender immigrant Latinas, stage in emotional and/or physical transition, gender identity, gender presentation, sexual orientation, age of migration, social class, bilingual abilities and coping styles are key factors to consider. The second system is the *microsystem*, which consists of a person's immediate environment, such as access to physical and mental health care and other social services. For transgender immigrant Latinas, this system includes important relationships that provide social, financial and/or facilitative supports such as family of origin as well as family of choice. Due to their multiple and intersecting identities, acculturative stress is also important to consider at the microsystemic level as transgender immigrant Latinas come into contact with a number of communities that include the LGBTQ, Latina/o and immigrant communities as well as 'majority' groups (i.e. heterosexual and white). The *mesosystem* focuses on the relations between different contexts, such as family and community. The quality of relationships with health care providers (e.g. physicians and psychotherapists) is especially important for transgender immigrant Latinas since many must rely on them for health care as well assessment during the asylum-seeking process. Furthermore, health care providers also assist women who are recovering from complex trauma as well as those initiating and/or continuing with transition. The *exosystem* incorporates social contexts and processes that affect transgender immigrant Latinas in ways they do not directly control. For example, women may be directly influenced by policies about immigration although they have not participated in creation of these policies or have the legal capacity to vote on them. In fact, transgender immigrant Latinas often have little to no power to influence government policies since many migrate to the US without legal documentation to work or reside in the US. Furthermore, many struggle with establishing legal documentation in the US since the designated gender on their native country's identification materials (male) is inconsistent with their gender identity (female). Lastly, the *macrosystem* consists of the overarching values that define societies and cultures. Currently, comprehensive immigration laws and proposed legislature heavily impact undocumented immigrants as well as asylum seekers.

A unique quality of Bronfenbrenner's approach is the attention given to bidirectionality, which seeks to understand the interactions between the environment and the individual (Bronfenbrenner 1979; Chronister, McWhirther & Kerewsky, 2004). According to Yakushko and Chronister (2005), "bidirectionality represents the influence that each ecological context exerts on the individual's development as well as the individual's power to effect change in an ecological context" (p. 293). For example, an immigrant transgender Latina from Central America is likely to experience acculturative stress while living in the US, which frees the individual to express her gender identity and establish social networks with other transgender groups (Espín, 1997; 1999; Greene, 1994). In addition to bidirectionality, the ecological model considers the significance of multiple identities as well as multiple sociocultural influences on individuals, such as societal attitudes concerning immigration and institutional policies – all of which may impact transgender immigrant Latinas and others who share similar intersecting identities (Cook, Heppner & O'Brien, 2004).

Application of ecological psychotherapy with a transgender immigrant Latina

Erika (pseudonym) was a research participant through a previous community-based qualitative study that examined transgender immigrant Latinas' mental health needs. In her native country of Honduras, Erika experienced discrimination and physical abuse from family and community members in relation to her gender identity, even escaping threats of being burned alive for being transgender. These experiences prompted Erika's migration to the US. Erika fled Honduras in her early 20s because she believed that she would not find employment or receive acceptance and support from family and community and because she believed that she would have to survive a lifetime of bias, discrimination and death threats if she continued living as female.

Without legal documentation to reside or work in the US, Erika struggled to find employment that prompted her engagement in high risk "survival work" to earn money. Erika expected to find work in a restaurant when she arrived in the US but had not yet secured employment after over two years in the country. Erika felt that her employment opportunities were especially compromised because of the ways her transgender identity overlapped with immigration; she did not have US documentation, and her identification from Honduras was not consistent with her female gender identity. Thus, Erika was left to seek employment while presenting as male because it matched her Honduran identification. Erika was often seeking employment in establishments whose tasks did not require US documentation (e.g. washing dishes and working in a flea market), but she was worried that presenting as female would "out" her as transgender and would complicate her ability to secure employment among Latina/o and/or immigrant establishments. Furthermore, she worried that the small, highly networked immigrant community's knowledge of her transgender identity would compromise her ability to seek future employment and access to critical resources, such as housing.

The challenges Erika experienced with securing employment prompted her involvement in sex work, which she described as detrimental to her mental health. Namely, Erika suffered several traumatic experiences during sex work that compromised her mental health and contributed to bouts of severe depression that included suicidal ideation and intent. Erika engaged in drug use as a means to cope with the traumatic experiences she faced, which eventually led her friends to actively encourage her involvement in a substance abuse recovery program. At Erika's lowest point, she described a strong desire to "not wake up", expressing hopelessness that resulted from the overlapping oppressions she experienced as a transgender immigrant Latina in the US.

Ecological psychotherapy recommendations at each level of the ecological model

Individual considerations

Many ecological factors affected Erika's individual experience in relation to larger social, economic, cultural and political systems. As a transgender immigrant Latina, Erika contended with individual encounters of transphobia, simultaneously negotiating ethnic discrimination, xenophobia, linguistic obstacles and legal limitations associated with immigration. Considering Erika's personal experiences over her lifespan, the social and cultural conditions she experienced growing up precluded her from expressing her female gender identity in Honduras, while in the US she continued to face challenges with living as a self-identified female. Specifically, Erika presented as male when seeking employment because it was consistent with her documentation and was her only means of securing employment and not outing" herself as transgender within the highly networked Latina/o immigrant community.

Recommended interventions for supporting Erika at the individual systems level include conducting an in-depth assessment of her current level of functioning, her access to socio-emotional support, her access to medical care, her ability to earn money and secure housing and her life history in Honduras and the US – up until her motivation and ability to seek out mental health support. It is important for mental health professionals to keep in mind that Erika's seeking support may have more to do with access (e.g. financial costs and documentation needed to receive services) and belief in a professional's ability to support her without bias than with sheer motivation to seek support. More generally, interventions at the individual systems level for transgender immigrant Latinas should involve use of a trained interpreter in behavioral health and immigration services; assessment of acculturation, immigration history and bilingual ability when determining mental health supports; thorough history of pre- and post-migration, including consideration of traumatic experiences that may have prompted migration and/or was experienced in route to the US (e.g. physical, emotional and sexual abuse); and identification of coping styles and strengths. Furthermore, if they are seeking asylum in the US based on transgender identity, then mental health and legal providers should be cognizant of how gender identity has impacted the client's life. For example, lack of employment opportunities related to transgender identity and immigration status may prompt survival work, such as sex work (Sausa, Keatley & Operario, 2007), that sometimes leads to a criminal record or contraction of a communicable disease that makes it harder for women to attain asylum (Immigration Equality Inc., 2012).

Microsystem considerations

Erika's well-being had been restricted by social barriers preventing her safe gender expression and transition. During childhood in Honduras, Erika suffered rejection and abuse by her family and neighborhood community. She longed for the support of close relationships. Her ability to participate in educational and professional opportunities was curbed by a combination of intense homophobic and transphobic bias and discrimination. In school, and in seeking employment, Erika refrained from expressing her female gender identity to avoid bias and discrimination.

Interventions at the microsystemic level could involve increasing Erika's support system, which may involve establishing relationships with LGBTQ, Latina/o and/or immigrant communities that reflect Erika's lived experiences and needs. Furthermore, mental health support from clinicians and social support agencies may help Erika develop coping skills in order for her to address past experiences of trauma as well as to learn healthy mechanisms for negotiating current and future challenges she may experience in relation to her transgender, immigrant and/or other cultural statuses. It is well known that transgender women of color face serious threats to their well-being on a regular basis that include verbal, physical and sexual assault, among other incidents of violence (National Coalition for Ant-Violence Programs, 2012). Mental health professionals should assist women to identify dangerous situations, to employ self-defense when faced in such a predicament and to know resources for legal protection, socio-emotional supports to cope with regular discriminatory experiences.

With respect to the world of work, Erika would also greatly benefit from vocational counseling to help determine her educational needs and occupational interests, with the intent of helping Erika find a path toward economic independence. Furthermore, vocational counselors and other help providers should provide Erika with information about her legal rights as a transgender immigrant woman in the event that she experiences discrimination from employers (see O'Neil, McWhirter & Cerezo, 2008).

Mesosystem considerations

The mesosystemic level of the ecological model involves understanding the relations between important individuals and/or systems in Erika's immediate environment (e.g. family of origin, peers and neighbors). An example of such a relation occurred when Erika received support from multiple facets of her social network during substance abuse recovery that facilitated her growth and development of effective coping skills. Namely, Erika received socio-emotional and financial support from friends, particularly other transgender Latinas, who encouraged her involvement in a substance abuse program. Friends not only offered continued encouragement throughout recovery but also provided financial help in the forms of calling cards and weekly allowance to prevent engagement in risky behaviors (e.g. sex work) to earn money.

The support from friends was fortified by staff in a social service agency that provided Erika with opportunities for leadership training and advocacy development. Erika's involvement with the agency reduced her engagement in sex work and increased her engagement in substance abuse recovery. Erika was hopeful about her ability to find work, recover from substance abuse and eventually attain documentation to reside and work in the US. Her hopeful demeanor and growing coping skills were a direct result of the combined socio-emotional support she received from peers as well as the socio-emotional and facilitative support afforded by the social service agency. Together, these supports fortified a positive, healthy path toward self-actualization.

Exosystem and macrosystem considerations

Factors within the exosystemic and macrosystemic levels significantly impacted Erika's quality of life. At the exosystemic level, Erika negotiated policies related to particular groups (e.g. transgender people and immigrants) that often did not consider the complex interactions of her multiple identities. For example, access to low-income housing was a challenge due to Erika's transgender identity (e.g. other residents being hostile toward her) while hormone therapy was impacted by her lack of documentation and limited English proficiency. Thus, Erika was constantly negotiating policies that represented only a portion of her identity but that were ultimately not reflective of her lived experience and consequent needs.

Policies within the exosystemic level are directly influenced by societal values present within the macrosystemic level, specifically messages about transgender identity, immigration, race, social class, and so on. Transgender immigrant Latinas often struggle with the constant threat of pervasive transphobia, xenophobia, racial profiling and threats of deportation. Together, these cultural values and messages affected Erika's options for housing, employment, healthcare, and so on, which were foundational to her quality of life and health outcomes.

Professionals who implement interventions at the macro- and exosystemic levels should be cognizant of both the client's ability to self-advocate as well as the responsibilities of psychologists to effect change in the interest of underserved communities. In regard to self-advocacy, finding support networks and community-based advocacy groups may also serve as protective factors to decrease self-destructive behaviors and increase positive perceptions of self. For example, Erika received leadership and advocacy training from a social service agency that can be used to change organizational culture for institutions that provide housing, health care, legal supports, and so on. Erika could provide testimony, or write a letter if lack of documentation is an issue, in front of a city's human rights commission and/or city commissioners who draft city and county policy and who often have power to direct the allocation of funds for public services. Mental health providers should encourage the use of self-advocacy that can be used in a public forum or in less formal ways that help women better understand the systems and barriers they negotiate on a daily basis. Ultimately, it is critical that women recognize their rights in relation to transgender and immigrant status and that they are able to self-advocate when their rights are infringed upon.

Recommendations for professional development

As the Latina/o population in the US continues to grow, it is likely that mental health professionals will experience contact with transgender immigrant women at some point in their careers. Therefore, it is imperative that mental health professionals be equipped with the awareness, knowledge and skills to effectively work with and advocate for this unique, underserved population. Mental health professionals may educate themselves by reading articles or books and attending conferences that focus on transgender and immigration issues to better inform themselves regarding knowledge content. National organizations that work on transgender rights include the National Center for Transgender Equality, the Transgender Law Center, Lambda Legal and The Sylvia Rivera Law Project, among others. An important consideration is that most of these organizations are staffed by transgender individuals and are at the cusp of the movement for equality, while some have specific policies on working with transgender immigrants. Mental health professionals should take it upon themselves to become familiar with national and local organizations working in the interests of transgender and immigrant communities as a means to build knowledge and skills that are current and reflective of the transgender and immigrant communities.

Self-reflective exercises are essential for mental health professionals to examine their own attitudes and potential biases that may affect the clinical relationship and influence their role as advocates. Also, it is important to remain informed and sensitive toward the use of various pronouns that are appropriate for a person's gender identity and to advocate for more gender-inclusive terminology. Pronoun use (e.g. she, he, hir, ze and they [gender neutral]) when communicating with and about a transgender client is imperative for affirming the client's gender identity because it expresses adherence to using language that is consistent with the client's gender self-identification. Within training settings, students and clinical training directors should locate and/or help develop the capacity for community-based practicum sites that serve transgender as well as immigrant communities that cover a range of needs, spanning from intake forms to termination plans. Such practices will improve women's experiences with a practicum site, which in turn may result in increased retention in clinical services that benefit both the client and clinicians-in-training.

Seeking consultation and supervision with experts who have experience with transgender immigrants is imperative; increased knowledge and understanding will allow for critical, difficult dialogues and will build competency across the disciple to best support this community. Individuals may also serve as advocates for this particular group by building partnerships and alliances with immigration lawyers, domestic violence shelters and other community organizations that focus on LGBTQ issues and immigration. Furthermore, training programs should take stock of their student body and take steps toward creating paths for LGBTQ immigrant students to become key players in the discipline. The mere act of advocating for transgender Latinas is not enough – we must create space for their voices within the discipline so that trends in the field reflect transgender immigrant Latinas' lived experiences.

Mental health professionals' own efforts at systemic change may include involvement in non-profit service organizations that direct services to underserved communities, such as joining boards of advisors, providing clinical consultation to staff members and/or contributing to grant writing to secure funds to maintain and strengthen the agency's efforts. Mental health professionals can also engage in change efforts within the larger mental health field, such as drafting aspirational guidelines and/or policy briefs in the interest of transgender and/or immigrant communities that are discipline-wide. For example, the American Counseling Association has adopted competency guidelines for providing mental health support to transgender and gender nonconforming clients (ACA, 2010) that spans professional domains of clinical services to research. The authors, in the first competency area of "human growth and development", write that competent counselors "affirm that all persons have the potential to live full functioning and emotionally healthy lives throughout their lifespan while embracing the full spectrum of gender identity expression, gender presentation, and gender diversity beyond the male-female binary" (p. 6), thus pointing out, from the outset of the competency guidelines, that transgender people deserve the right to be respected and affirmed in their gender identity by mental health professionals.

Conclusion

The purpose of this chapter was to illuminate the life of Erika, a transgender woman who immigrated to the US from Latin America. We set out to complete this task with the goal of informing mental health professionals about sociocultural factors impacting many women of this community, oftentimes leading to severe marginalization and psychological trauma that calls for culturally informed mental health care. Given the various factors, including discriminatory treatment and policies that commonly underserve transgender and immigrant communities, we used an ecologically based theoretical model that allowed for the consideration of various ecological spheres impacting Erika's life. Although Erika's life story is unique to her, the challenges she faced as well as the community resources and supports that enabled her to move closer to her life goals provide a model for how mental health professionals can better support and advocate in the interest of transgender immigrant women.

References

American Counseling Association. (2010). Competencies for counseling with transgender clients. *Journal of LGBT Issues in Counseling, 4*, pp. 135–139.

American Psychological Association (2011). *Guidelines for psychological practice with lesbian, gay, and bisexual clients*. Washington, DC: American Psychological Association.

Anderson, L. (2010). Punishing the innocent: How the classification of male-to-femaletransgender individuals in immigration detention constitutes illegal punishment under the fifth amendment. *Berkeley Journal of Gender, Law & Justice, 25*(1), pp. 1–31.

Association for Lesbian, Gay, Bisexual, Transgender Issues in Counseling (2009). *Competencies for Counseling with Transgender Clients*. Alexandria, VA: American Counseling Association.

Bronfenbrenner. U. (1979). *The ecology of human development*. Cambridge, MA: Harvard University Press.

Cantú, L. (2009). *The sexuality of migration: Border crossings and Mexican immigrant men*. New York: New York University Press.

Center for American Progress (2013). Retrieved from http://www.americanprogress.org/issues/immigration/report/2013/04/03/59040/the-facts-on-immigration-today-3/#population (July 14, 2013).

Center for Disease Control (2008). *HIV/AIDS Surveillance Report, 2006, 2008*. Retrieved from http://www.cdc.gov/mmwr/preview/mmwrhtml/mm5640a4.htm?s_cid=mm5640a4_e (July 14, 2013).

Cerezo, A., O'Neil, M.E. & McWhirter, B.T. (2009). Counseling Latina/o students from an ecological perspective: Working with Peter. *Journal of College Counseling, 12*, pp. 170–181.

Chávez, K. R. (2010). Border (in) securities: Normative and differential belonging in LGBTQ and immigrant rights discourse. *Communication and Critical/Cultural Studies, 7*(2), pp. 136–155.

Conyne, R.K. & Cook, E.P. (2004). *Ecological counseling an innovative approach to conceptualizing person-environment interaction.* Alexandria, VA: American Counseling Association.

Cook, E.P., O'Brien, K.M. & Heppner, M.J. (2004). Career counseling from an ecological perspective. In R.K. Conyne & E.P. Cook (Eds.), *Ecological counseling an innovative approach to conceptualizing person-environment interaction* (pp. 219–241). Alexandria, VA: American Counseling Association.

Chronister, K.M., McWhirter, B.T. & Kerewsky, S.D. (2004). Counseling and ecological prevention practice. In R.K. Conyne & E.P. Cook (Eds.), *Ecological counseling: An innovative approach to conceptualizing person-environment interaction*, (pp. 315–338). Alexandria, VA: American Counseling Association.

Crowther, J.R. (Ed.) (1995). *ELISA: theory and practice, 42*. Springer.

Espín, O.M. (1997). Crossing borders and boundaries: The lives of narratives of immigrant lesbians. In B. Greene (Ed.), *Ethnic and cultural diversity among lesbians and gay men* (pp. 191–215). Thousand Oaks, CA: Sage Publications.

Espín, O.M. (1999). *Women Crossing Boundaries a psychology of immigration and transformations of sexuality.* New York: Routledge.

Gainor, K.A. (2000). Including transgender issues in lesbian, gay and bisexual psychology. *Education, research, and practice in lesbian, gay, bisexual, and transgendered psychology: A resource manual, 5*, pp. 131–160.

Greene, B. (1994). Lesbian women of color: Triple jeopardy. In L. Comas-Diaz & B. Greene (Eds.), *Women of Color: Integrating ethnic and gender identities in psychotherapy* (pp. 389–427). New York: Guildford Press.

Huffington Post (2012). *Joe Biden: Transgender discrimination is 'The Civil Rights Issue Of Our Time.'*

Immigration Equality Inc. (2012). Retrieved from http://immigrationequality.org/issues/transgender/ (August 31, 2013).

Morales, A., Corbin-Gutierrez, E.E. & Wang, S.C. (2013). Latino, immigrant, and gay: A qualitative study of their adaptation and transition. *Journal of LGBT Issues in Counseling, 7*, pp. 125–142.

Morales, E. (2013). Latino lesbian, gay, bisexual, and transgender immigrants in the United States. *Journal of LGBT Issues in Counseling, 7*, pp. 172–184.

O'Neil, M.E., McWhirter, E.H. & Cerezo, A. (2008). Transgender Identities and Gender Variance in Vocational Psychology Recommendations for Practice, Social Advocacy, and Research. *Journal of Career Development, 34*, pp. 286–308.

Pew Hispanic Center (2012). Retrieved from http://www.pewhispanic.org/2012/12/06/unauthorized-immigrants-11-1-million-in-2011/ (July 14, 2013).

Reynolds, V., Falger, V. & Vine, I. (1987). The Sociobiology of ethnocentrism: evolutionary dimensions of xenophobia, discrimination, racism, and nationalism. Athens, GA: University of Georgia Press.

Ritter, K.Y. & Terndrup, A.I. (2002). *Handbook of affirmative psychotherapy with lesbians and gay men.* New York: Guildford Press.

Sangganjanavanich, V.F. & Cavazos JR, J. (2010). Workplace aggression: Toward social justice and advocacy in counseling for transgender individuals. *Journal of LGBT Issues in Counseling, 4*, pp. 187–201.

Santiago-Rivera, A.L., Arredondo, P. & Gallardo-Cooper, M. (2002). *Counseling Latinos and la familia*. Sherman Oaks, CA: Sage Publications.

Seuffert, N. (2009). Reflections on transgender immigration. *Griffith Law Review 18*(3), 428–452.

Sausa, L.A., Keatley, J. & Operario, D. (2007). Perceived risks and benefits of sex work among transgender women of color in San Francisco. *Archives of Sexual Behavior, 36*, pp. 768–777.

Suarez-Orozco, C. & Suarez-Orozco, M.M. (2001). *Children of immigration*. Cambridge, MA: Harvard University Press.

Suarez-Orozco, C., Suarez-Orozco, M. M. & Tordova, I. (2008). *Learning in a new land: Immigrant students in American society*. Cambridge, MA: Harvard University Press.

National Center for Transgender Equality and the National Gay and Lesbian Task Force (2011). *Injustice at every turn*. Washington, DC: Author.

National Coalition of Anti-Violence Programs (2012). Retrieved from http://www.avp.org/storage/documents/ncavp_2012_hvreport_final.pdf (August 31, 2013).

Transgender Europe (2009). *Trans murder monitoring project*. Retrieved from http://www.transrespect-transphobia.org/en/tvt-project/tmm-results/liminalis-2009.htm (July 14, 2013).

Yakushko, O.F., Watson, M. & Thompson, S. (2008). Stress and Coping in the lives of recent immigrants and refugees: Consideration for counseling. *International Journal for the Advancement of Counseling, 30*, pp. 167–178.

Yakushko, O.F., Backhaus, A., Watson, M., Ngaruiya, K. & Gonzales, J. (2008). Career development concerns of recent immigrants and refugees. *Journal of Career Development, 34*, pp. 362–396.

Yakushko, O.F. & Chronister, K.M. (2005). Immigrant women and counseling: The invisible others. *Journal of Counseling and Development, 83*, pp. 292–298.

U.S. Census Bureau (2011). *Educational attainment of the population by 25 years and over by sex and Hispanic origin type: 2009. Current population survey, annual social and economic supplement, 2009*. Retrieved from http://www.census.gov/population/www/socdemo/hispanic/cps2009.html (July 14, 2013).

9

A Study of Immigrant Latina Survivors of Domestic Violence: Becoming Líderes Comunitarias (Community Leaders)

Josephine V. Serrata, R. Lillianne Macias, Alvina Rosales, Rebecca Rodriguez and Julia L. Perilla

Domestic violence (DV) is a global human rights issue affecting all women regardless of race or culture (Tjaden & Thoennes, 2000). Immigrant Latina survivors of DV represent a heterogeneous group that experience not only violence in the home but also multiple ecological stressors deeply influenced by oppressive forces, including gender (Pessar & Mahler, 2003). Because partner violence almost never occurs in isolation, many immigrant Latina survivors simultaneously cope with difficulties arising from sexism, acculturation, racism, discrimination, immigration enforcement policies and limited access to resources. Moreover, immigrant Latinas exist within gendered systems of power that influence and interact with intersecting identities of sexuality, nationality, social class and more (Pessar & Mahler, 2003). Notably, immigrant Latina survivors often demonstrate significant strength and fortitude in coping with tremendous stress in which partner violence is only one aspect of their lives (Kofman, Phizacklea, Raghuram & Sales, 2000; Perilla, Serrata, Weinberg & Lippy, 2012). The purpose of this study is to extend research on immigrant Latina/o communities and domestic violence by highlighting the strengths and leadership abilities of a group of immigrant Latina survivors of DV.

While deficit-focused research and interventions continue to dominate the social sciences, researchers have begun to document the significant strengths displayed by immigrant Latinas (e.g. Campbell, 2008; Perilla et al., 2012). Immigrant Latinas have the potential to draw on both personal and cultural strengths to cope with and prevent further violence in the face of multiple forms of oppression, including those due to immigration experience and status (Espín, 2010). By moving beyond deficit-oriented understandings of violence in immigrant populations that suggest that Latinas are merely passive in their immigration story (Kofman et al., 2000), we sought to explore the transformative process that occurs when immigrant Latina survivors of DV participate in an intervention to develop their own innate leadership abilities and become *Líderes (community leaders).*

Personal transformation is a powerful process, especially as individuals begin to see their experience as it relates to others in their immediate social networks, their community and other sociopolitical contexts (Prilleltensky, 2012). Transformation is especially relevant and important for understanding immigration and gender structures, and it always occurs at the intersections of sociopolitical forces such as immigration politics and violence, inclusive of personal histories of trauma and violence experienced in survivors' country of origin (Hondagneu-Sotelo, 2003; Pessar & Mahler, 2003; Zarza & Adler, 2008). The prevalence of domestic violence, for example, is not unique to immigrant communities or American culture, and so, women leaving their countries of birth may have existing experiences of violence at home (Tjaden & Thoennes, 2000; Zarza & Adler, 2008). For example, women have reported violence from a partner as a reason for immigrating (Zarza & Adler, 2008). Additionally, political violence may be part of Latina/o immigrants' experiences before entering the US, as in the experience of a Salvadorian immigrant that comes to the US but is not granted asylum (Portes & Rumbaut, 1996). The study presented in this chapter documents the process for a group of *Líderes* and an aspect of immigration often excluded from literature: giving voice to the transformative process in the development of women's strengths.

Immigrant Latinas & domestic violence

Various studies have highlighted the experiences of immigrant Latinas when they face DV, including culture-specific dynamics with partners and barriers to services (Brabeck & Guzman, 2008; Dutton, Orloff & Hass, 2000; Kelly, 2009; Moracco, Hilton, Hodges & Frasier, 2005; Vidales, 2010; Zarza & Adler, 2008). Thus, ecological and gendered frameworks are critical lenses for understanding the realities of immigrant survivors. Immigration status has been found to be a common and powerful control mechanism used by partners of immigrant women to force them to stay in the relationship (Dutton et al., 2000). Moreover, the origin of their DV experience sometimes predates their arrival in this country. For example, Zarza & Adler (2008) found that approximately 4 percent of participants had fled their countries of origin in search of protection and safety from DV. The contexts of both home and host countries are inseparable from power gender relations and roles that permeate across culture and national boundaries (Espín, 1999). Scholars on gendered migration have theorized that gendered social roles may transform as a result of migration. Upon moving to the US, some women experience an increase in their social standing due to the greater availability of jobs for immigrant women, while men may experience a decline in their social status in contrast to social roles in their communities of origin (Hondagneu-Sotelo, 2003). As a result, tensions may arise in the household of heterosexual Latina/o couples (Espín, 2010).

Potential benefits resulting from migration notwithstanding, Latinas affected by DV often face barriers in becoming more economically independent or are at greater risk for violence due to their new-found independence (Espín, 2010). For example, perpetrators have prevented women from obtaining a driver's license, deliberately delayed legal processes for immigrant women to work legally and interfered with childcare arrangements (Mankowski, Galvez & Glass, 2011). Unique forms of violence experienced by survivors of DV include threats by perpetrators to take away their children, threats to report partners to immigration authorities and even sending women to their country of origin temporarily as a control tactic (Kelly, 2009). Additionally, immigrant Latinas may face unique barriers to accessing services, possibly due to fear of deportation while seeking help from social services (Dutton et al., 2000). In states where harsh immigration laws have been enacted, immigrant Latina survivors report a decrease in the likelihood of calling the police due to heightened immigration enforcement policies and increased fear of deportation (Rodriguez, Serrata, Perilla & Macias, 2012). This literature offers important insight into the gender dynamics and unique problems relative to DV in Latina/o immigrant communities. Still, more studies are needed that examine the transformative experiences of immigrant women and the potential sources of strength and resilience that we have observed firsthand among immigrant survivors of DV (Espín, 1999).

Immigrant Latinas in Georgia

The 2010 US Census indicates that 10 percent of Georgia residents are Hispanic (Atlanta, Sandy Springs, Marietta, GA Profile Summary) and that this population is in a period of marked growth. Specifically, Atlanta's Hispanic population grew from 24,700 in 1980 to 270,400 in 2000 and to 547,400 in 2010, a growth rate over the last decade that was four times that of the Atlanta general population (102 percent vs. 24 percent increase, respectively). The steep growth rate for Latina/os in Atlanta has occurred at the same time as the Southeastern US is experiencing critical shortages of culturally appropriate, family-oriented medical and mental health interventions (Odem & Lacy, 2009). Moreover, Latina/os in Georgia are subject to intense discrimination based on immigrant status after the enactment of the Georgia immigration law (House Bill 87) passed in 2011. Similar to the now notorious law Arizona passed authorizing police to demand proof of legal residency for minor infractions such as moving traffic violations, this and similar laws have led to increased deportations across the country (Homeland Security Office of Immigration Statistics, 2011), resulting in an increased sense of fear shared by entire communities regardless of documentation status (Rodriguez et al., 2012).

Historically, new immigrants have relied heavily on extended family already settled in the host country for social and instrumental support. However, first-generation Latina/o immigrants are often at a disadvantage in learning how to navigate social services and educational systems because their extended families are scattered across the United States (Taylor, Lopez, Passel & Motel, 2011). Absence of practical help and social isolation undermine the stability of Latina/o immigrant families and can fuel parental depression, anger and marital conflict (Martinez, McClure, Eddy & Wilson, 2011). Contextual factors regarding immigration and policy are key issues that need to be acknowledged and shared when working in a Latina/o immigrant community. The anti-immigration sentiment in many states has resulted in fewer resources for immigrant communities, and both prevention and intervention strategies must now incorporate elements that support community members' self-empowerment and ability to navigate and challenge discriminatory systems.

Transformation for immigrant Latina survivors of DV

Given the complex social and personal challenges that immigrant Latina survivors experience daily, it is imperative that approaches to DV address the oppressive structures that sustain sexism, racism and violence. One approach is grounding DV programs within a self-empowerment framework to promote critical consciousness, or *concientización* (Freire, 1997; Martín-Baró, 1996). Similar to the feminist pedagogical strategy of "consciousness raising" (Bricker-Jenkins & Hooyerman, 1986) and to the feminist theoretical stance of "the personal is political" (Hanisch, 1969), *concientización* can be defined as the process of becoming aware of one's place in society and developing a critical understanding of the history of societal oppression and active opposition to this oppression as it exists in current society (Perilla, 1999). Immigrant Latinas must understand not only the history and current social conditions of their country of residence but also those they experienced in their country of origin. *Concientización* reminds women that they are not facing these life challenges alone. This concept is even more powerful when immigrant women are able to see that the answer to oppression is to work together to change the society in which they and their families live (Perilla, 1999). *Concientización* is a foundational element of the *Líderes* program, which is grounded in an empowerment process whereby participants learn to challenge oppressive forces by discovering their natural leadership abilities individually and collectively. The study presented in this chapter described the transformative process for the immigrant Latina survivors of DV who took part in the *Líderes* program.

Method

Participants

Training participants were recruited from Caminar Latino, a community-based organization in Atlanta, which has worked for over 23 years with Latina/o families affected by DV. The average age of the nine training participants was 36 and the average number of years of completed education was 11. All of the participants were born outside of the United States; six were from Mexico and one each from Colombia, El Salvador and Honduras. Eight had children, with an average of two children per trainee. All participants were employed and occupations included housekeeper, nanny, restaurant employee, pre-school teacher and hair stylist. On average, participants attended Caminar Latino for 3.6 years, with the shortest attendance time being six months and the longest being 11 years. At the time of the intervention, three participants regularly attended the Caminar Latino weekly support groups.

The *Líderes* program training sessions were facilitated by two women from Caminar Latino, including one who was also a graduate level researcher at Georgia State University. At the time of the *Líderes* training, both women worked closely together as facilitators in the women's support group at Caminar Latino (to preserve their privacy, pseudonyms will be used). Marisa, age 49, is a Mexican immigrant who has resided in the US since 1994 and worked as a women's advocate at Caminar

Latino since 2005. At the time of this study, Graciela, a 29-year-old Latina of mixed heritage (Panamanian and Mexican-American) volunteered at Caminar Latino as a researcher and student clinician for over four years.

The *Líderes* program is based on a self-empowerment model that explores and develops participants' natural leadership abilities. The program was originally developed by Casa de Esperanza, a nationally recognized organization originally established as a shelter in Minnesota by a small group of Latina advocates (Casa de Esperanza, 2010). Through collaborative efforts, the *Líderes* program was adapted and implemented at Caminar Latino in the spring of 2011. The program is meant to facilitate an individual and collective transformation so that at the end of the program attendees see themselves together with their colleagues as agents of community change. The women are trained in topics of community leadership and presentation skills as one of their primary mechanisms for change is through community *talleres* (workshops) where they present on topics that are relevant to their peers (e.g. parenting, bullying, immigration policies, safety planning, domestic violence, etc.). They are also trained as DV first responders so that they can provide support and resources if a workshop attendee or peer were to want to discuss DV.

Procedure

This study was conducted as part of a larger evaluation of the study's impact on training participants' sense of leadership capacity, leadership skill development, knowledge regarding community leadership and sense of connection to their cohort and organization. The research protocol was approved by the Institutional Review Board of Georgia State University. In answering the research question, *how do immigrant Latina survivors of DV experience the Líderes program?*, descriptions of the nine immigrant Latina trainees' experiences were collected using journals kept throughout the six-week program. The journals were distributed at the first session, and participants were informed that, if they agreed, the journals would be collected at the end of the intervention for the purposes of the study. Participants were prompted at the end of each session to write reflections in their journals with the following statement: *Remember, we are interested in learning about your experience with the program. Please take some time to write your reflections about your experience before the next session.* While all turned in a final reflective entry about their experience in the program, only four of the nine training participants provided all journal entries for the analysis.

One-on-one interviews lasting approximately 20 minutes were conducted in private offices at Caminar Latino with the two training facilitators. Both interviews began with asking the facilitator, *tell me about your experience adapting the Líderes program*, and prompts (e.g. *tell me more about what that was like for you*) were used to clarify or gather more detail on the experience of participants throughout the training period. Training agendas were used as visual cues to facilitate discussion of various points in the training process, and facilitators were asked generally to talk about the sessions (e.g. *tell me about your experience on this day.*).

Study design

This phenomenological study was part of a broader cultural adaptation and evaluation of the *Líderes* program (for a description of the program, see Serrata, 2012; Macias, 2013). Phenomenology is a theoretical framework that focuses on the meaning of subjective experience and how individuals make sense of the world (Kuper, Reeves & Levinson, 2008). Qualitative phenomenology guided the exploration of immigrant Latina survivors' lived, transformative experience of training for roles as leaders (in the larger community) and experiences of the Latina facilitators working closely with them as this process unfolded. The traditional separation between subject and researcher is challenged in qualitative theory (Guillemin & Gilliam, 2004). Thus, as Latina researchers, facilitators and in many cases survivors of domestic violence, we acknowledged and strived to bracket our own feelings and beliefs to listen to study participants.

Data analysis

Diaries were analyzed by the first author, who is bilingual. Interview transcripts were transcribed and Spanish text was translated into English by a translator outside of the research group. Each transcript was first read in its entirety in order to become familiar with the content (Riemen, 1986). The first and second authors highlighted and coded a list of significant statements (a technique known as horizontalization of the data; Moustakas, 1994), treating each separate non-overlapping, non-repetitive statement as having equal worth. The qualitative data analysis software NVivo 9 (QSR International) was used to assist with this step and the final synthesis of meanings to create a set of the salient themes. Pseudonyms are used to present significant quotes and examples of themes in this chapter in order to protect participants' privacy.

Results

Themes emerging from participant journals

To answer the research question, *how do immigrant Latina survivors of DV experience the Líderes program?*, thematic descriptions of the participants' experiences of the program – including the environment in which the program was delivered, personal transformation, and personal commitment to community – were extracted from their journals.

Program environment. Participants described experiencing the intervention through several activities and learning experiences. Program components that were discussed in journals included early activities that introduced participants and facilitators to one another, naming the group, understanding leadership, reflections on personal leadership styles and the development of presentation skills. Some activities were more salient for some members than they were for others, depending on how the activity resonated with their own values or experiences. For instance, one participant (Lourdes) discussed a key program element of developing presentation skills in every journal. From early on in the program this aspect of the curriculum was a source of significant anxiety for her:

> We just got out. They explained to us our responsibilities and expectations. This looks more difficult than I had expected. I think that I will not be able to do it. It includes things that I am very afraid of: speaking in public.

Given each participants' experience with violence, it is not surprising that particular aspects of the program could be anxiety provoking as the consequences of DV span across many areas of life (Aldarondo, Kaufman Kantor & Jasinski, 2002; Bonomi, Anderson, Cannon, Slesnick & Rodriguez, 2009) and were expected to emerge during the program (components of the curriculum were adapted to address coping with stress and triggers).

Nevertheless, each participant wrote about having overall positive reactions to the intervention. Specifically, experiences with the leadership intervention can be understood within the broader context of supportive relationships. Two participants described difficult obstacles early on, and they expressed feelings of nervousness and anxiety about their own abilities. One participant described difficulties connecting with the other group members and considered quitting the program. In speaking with another group member (another Líder), she was able to establish a supportive base and was encouraged to speak with the group facilitators. The support from the other group member was important for her continuing with the group. Similarly, Lourdes wrote about her doubts that she had the necessary tools to conduct presentations; however, as the program progressed, her confidence improved within the context of a secure and supportive learning environment.

> She [the facilitator] explained some points [about the presentation] that I did not understand. Now, I feel a little more secure [about presenting].

As noted by previous scholars (Hughey, Peterson, Lowe & Oprescu, 2008; Speer, Jackson & Peterson, 2001), the supportive environment of the program was a key component for the participants to feel connected and thus willing to partake in a self-exploration process. In addition, a supportive environment may be especially important for survivors of violence; one scholar has suggested that "groups provide the possibility not only of mutually rewarding relationships but also of collective empowerment" (Herman, 1997, p. 216).

Personal transformation. A key element of an empowerment process is the acquisition of power and control over a bothersome issue (Rappaport, 1987). Interestingly, three of the four women described a similar process of transformation in their journals regarding their sense of leadership. In the first journal entries for each participant the women described difficult feelings such as nervousness, not feeling competent and not identifying themselves as leaders. For example, Lourdes noted the following:

> Today, I do think that I am going to quit because this feels like it's not for me. I see my colleagues [as] confident and secure, and I feel lost.

This statement reflects the participant's thoughts about leaving the group and feelings of insecurity about her own abilities. However, in a later journal entry, Lourdes noted a change in her view of herself as a leader:

> Today something happened to me that was really great. I feel more confident to do a presentation. I impressed myself with what I could do today.

The transformation process of a participant who in early sessions did not feel competent in her abilities is clearly reflected here. As the sessions progressed, she shifted the perspective she had of herself as being more competent and able to be a leader in her community. This theme is reflected in her later journals. Another member with previous experience in leadership training described the process of rediscovering her abilities and remembering that she had existing skills that would benefit her in her new leadership role. All women described a process of transformation as a key element in their experience of the program.

Motivation to impact the community. A unique aspect of the empowerment process is that individual change is embedded within interactions of the individual with their community and other sociopolitical forces (Zimmerman, 1995). This process is evident in the theme of sense of community connectedness that emerged from the data analysis. Although the program was difficult and anxiety provoking for many of the participants for different reasons (e.g. for some it was the thought of presenting and for others it was the idea of facilitating conversations), they all described feeling motivated to continue due to their connection to the community. They also described completing the program and developing their skills so they could give back to their communities. The prayer that Consuelo wrote in her journal exemplifies this quite well:

> Dear God, my desire/hope is to have the wisdom, enough love to listen, support, share, and comfort my peers.... Do not leave me alone, Lord, and provide me the desire to serve my community, I beg you, Lord, in the name of your beloved son Jesus, hear my prayer.

This prayer reflects this participant's commitment to impacting her community. Part of this participant's experience is drawing on her spirituality as strength to help her influence her community peers. Another participant wrote in her journal the importance of using what she learned to influence and support others in a positive manner. It appears as though an important element of this experience for the participants was the focus on community involvement and obtaining skills that they could use to give back to their communities.

Themes emerging from the facilitator interviews

Impact of histories of violence. Thematic coding of the interviews of facilitators revealed an understanding of the role of trauma in the personal histories of the participants. For one of the facilitators, Marisa, the shared experience of being an immigrant, a Latina and a survivor of domestic violence was acknowledged early on:

> We were all immigrants for the first time, yes, and we are all survivors. So I think that is what unites us more. And all of us had the spirit or desire to help someone else, right? So I think that's what, that's what has kept us tied together.

The facilitators' narratives revealed the importance of considering the participants' violence and trauma history. The *Líderes* program at Caminar Latino was the first to be implemented with an entire group of survivors of domestic violence. Facilitators described the collaborative decision with trainees to add exercises and activities to manage anxiety triggers and difficult content. Here is what Graciela, one of the facilitators, had to say:

> They talked about how they needed skills for when their nervousness was triggered in group [meetings], and it was really hard to control. And I had to stand back and understand what they were asking for – ways to handle triggers. As survivors of so much trauma...it can trigger all those emotions from your trauma experience as well. They are so closely related, so for them being part of the group and being nervous, it was a lot more emotions than for someone who is just dealing with public speaking, because it was so related to their stories.

Personal transformation. Similar to the transformation described by the participants, both facilitators described witnessing transformation as highlighted by scholars of the empowerment process (Zimmerman, 1995; Rappaport, 1987). Both facilitators discussed their observations of training participants' initial discomfort and feelings of anxiety, particularly as they were challenged to use new leadership skills. Marisa stated:

> [For] most of them, it was fear that they were being watched. They felt nervous. They felt a lot of pressure that everyone was watching them, right? They all got nervous [and] then one [participant] didn't get up in front [to present].

Despite these mixed feelings of anticipation in the early stage of the program, the facilitators' narratives also reflected a positive experience of transformation, both personal and also observed in their work with the *Líderes* trainees. Both facilitators' observations reflected the transformative experience of training participants. Marisa said:

> And now I see that, in the beginning the feeling [from the participants] that 'I can't do it' – the feeling 'no, this isn't for me, this is for another who truly will be a leader, and not me.' And to see them now so involved, so ready to give to the community, bringing up subjects/issues that I don't know at all, that I have no information for...and that they want to learn more, their concern to know more about computation/computers, their concern to know more things in order to have more tools to continue with their workshops. They can do positive things, first for themselves and then for the community, right? – which in the end, affects us all.

Trauma theorists have put forth social participation as a means of resolution and development of survivors' existing strengths, something Herman suggests can "transform the meaning of their personal tragedy by making it the basis for social action" (Herman, 1997, p. 207). More research is needed with larger groups to support the link between recovery and social action; however, the observations of the training facilitators in this study reflect the importance of commonality for survivors and the potential for leadership development programs as safe places to process triggers and develop skills for connecting to a broader mission to end violence.

Discussion

This study offers a different perspective of Latina immigrants, one that highlights personal transformation within the context of their challenging life realities. In this study, participants and facilitators of the *Líderes* program identified the importance of acknowledging histories of violence, of a supportive environment from which the program was delivered, of personal and observed discovery of natural leadership abilities and of a collective commitment to creating community change. These findings highlight how immigrant Latinas, and particularly those whose lives have been touched by violence, carry with them a personal knowledge of energy and life *beyond* (rather than in spite of) great hardship. Moreover, the women in this study challenge the misconception that immigrant women are passive participants in the immigration experience (Kofman et al., 2000), showing us instead their eagerness to become active agents in their communities.

The influence of a gendered structure and process are inherent qualities of the experiences and themes uncovered in this study. Leadership can be considered within the public sphere so often ascribed to men, and so, the apprehensions expressed by women in this study can be viewed in the context of a challenge to these norms (Mahler & Pessar, 2006). Given these findings, the attention to gender also highlights the arduous and sometimes contradictory realities for Latina immigrants. Immigrant women living in the United States without the privilege of citizenship are often invisible and historically excluded from ordinary, day-to-day life, and in particular, they are not considered as potential leaders of community enhancement and social change. This is especially true when anti-immigrant policies are a part of their larger environment. At the same time, Latinas potentially may have more "gains" than men from fluid gender expectations as a result of migration and employment (Mahler & Pessar, 2006). This context is an important lens for understanding the self-doubt first expressed by the women as they were beginning the program, which is grounded in challenging social norms.

Interestingly, this study illustrates positive transformation from doubt and apprehension about their ability to be community leaders to the discovery of their own innate power and potential as *Líderes*, highlighting the purposeful decisions of the women to become agents of change that transcend securing personal and family well-being. Their commitment to community change also appeared to transcend their initial apprehensions. Findings reflect the increased feelings of self-confidence to be community leaders and participants' acknowledgment of the potential to impact their communities, despite their intersecting experiences with oppression, particularly violence. This transformation is especially salient given the concurrent passing of anti-immigrant legislation in Georgia while the women were participating in this program.

As Prilleltensky (2012) noted, personal transformation is most powerful when the person sees her/his experience as it relates to others in their community and other socio political contexts. This study complements his findings and those of other scholars who have noted the importance of the collective experience, especially when working with women of color (Zimmerman, 1995; Worell & Remer, 2003). In this study, the experience of a supportive environment was critical to personal transformation, and for some women, the supportive environment determined their continuation in the program. This finding highlights the importance of the organization's ability to implement a program grounded in the value of self-empowerment and mutual support as recommended by feminist tenants. In addition, the women's collective commitment to give back to their own communities supports the self-empowerment process proposed by Rappaport (1987). The program facilitated the acquisition of mastery and control, and it promoted a sense of confidence that the group, as a team, can challenge social norms.

The consideration of trauma histories was particularly salient for the facilitators of the program. Because the *Líderes* program at Caminar Latino was implemented for the first time with survivors of DV, the facilitators' narratives revealed personal trauma histories as being a salient aspect of the participants' transformative experience. Marisa, a facilitator, observed during the interview that all the women connected to the program shared a common experience of domestic violence that contributed to the cohesion and "spirit" of the group. The facilitators' ability to connect with the participants and understand their lived experiences of immigration and violence included responses to requests to adapt the leadership training program in order to address possible trauma triggers as related to their presenting and talking about the topic of DV, including relaxation exercises. It is important to note that Latina immigrant survivors of DV may carry the burden of not only significant interpersonal and physical trauma in the United States but also past experiences of violence, including domestic, political and social violence in their countries of origin (Zarza & Adler, 2008) as well as tensions due to changing gender roles (Espín, 1999). The ability of researchers and practitioners to listen to and understand the experiences of immigration and trauma may prove to be critical in our work not just with Latina immigrants touched by violence but also with their families and communities.

Implications for practice with this population include the benefit and need for the development of strengths-focused practice and reframing of static conceptualizations of sexuality and gender in the context of DV. As highlighted in the findings, for immigrant Latinas it is imperative that programs not only consider personal histories of violence but incorporate a deep understanding of gendered experiences carried over from their countries of origin or during the immigration process. Organizations, advocates and professionals must recognize contextual challenges with which immigrant Latinas contend on a daily basis, including communities with aggressive and repressive policies against recent immigrants. Additionally, this study highlights the importance of the collective experience of the participants and the environment in which the program is delivered. Thus, programs that are considering implementing a similar approach should carefully assess their readiness and infrastructure for supporting an empowerment-based program.

Future studies must continue to document outcomes of interventions that highlight strengths and transformative processes that might not surface in traditional DV services. Further, a focus on the role of gender necessitates the inclusion of men in research and discourse on sexuality in the context of DV, as well as studies understanding DV among same-sex partners and sexual minorities (Pessar & Mahler, 2003). In addition, future research that explores the effectiveness of the *Líderes* program as it is implemented in other Latina/o communities will enhance our understanding of transformation and leadership emergence in other settings and populations, including migrant camps, rural areas and other diverse Latino communities.

This article documents the process of transformation for a group of immigrant women survivors from multiple and diverse Latin American countries who now actively work as leaders in their communities. Since the training of the first cohort of *Líderes* at Caminar Latino, the program has grown exponentially. On January 29, 2011, the *Líderes* held their first community workshop on domestic violence and also developed a workshop on bullying. Now six of the nine graduating *Líderes* meet monthly to plan community activities and have also expanded their roles within Caminar Latino, including raising funds to benefit the DV organization by organizing several large community yard sales and volunteering at the annual community Thanksgiving celebration. Additionally, the women have been featured on local radio and have provided community outreach at health fairs, churches and homes. The group continues to increase their own knowledge about topics relevant to the Latina/o community by inviting guest speakers from a range of organizations and professionals, including the Mexican Consulate, psychologists and immigration lawyers, to monthly meetings. One *Líder* has now become a facilitator for the domestic violence support group where she first began her journey as a survivor years ago. The *Líderes'* transformative stories revealed through this study deepen our understanding of the complex experience of Latina immigrant survivors.

References

Aldarondo, E., Kaufman Kantor, G. & Jasinski, J.L. (2002). A risk marker analysis of wife assault in Latino families. *Violence Against Women, 8*(4), pp. 439–454.

Atlanta, Sandy Springs, Marietta, GA Profile Summary (n.d.). In Diversity data project. Harvard School of Public Health. Retrieved from http://diversitydata.sph.harvard.edu/Data/Profiles/Show.aspx?loc=124

Bonomi, A.E., Anderson, M.L., Cannon, E.A., Slesnick, N. & Rodriguez, M.A. (2009). Intimate partner violence in Latina and Non-Latina women. *American Journal of Preventive Medicine, 36*(1), pp. 43–48.

Brabeck, K.M. & Guzman M.R. (2008). Frequency and perceived effectiveness of strategies to survive abuse employed by battered Mexican-origin women. *Violence Against Women, 14*(11), pp. 1274–1294.

Bricker-Jenkins, M. & Hooyman, N. (1986). Feminist pedagogy in education for social change. *Feminist Teacher, 2*(2), pp. 36–42.

Campbell, W.S. (2008). Lessons in resilience: Undocumented Mexican women in South Carolina. *Affilia: Journal of Women and Social Work, 23*(3), pp. 231–241.

Casa de Esperanza. (2010). The Latina peer education initiative. Casa de Esperanza: McNeely.

Dutton, M.A., Orloff, L. & Hass, G.A. (2000). Characteristics of help-seeking behaviors, resources, and services of battered immigrant Latinas: Legal and policy implications. *Georgetown Journal on Poverty Law and Policy, 7*(2), Pp. 245–303.

Espín, O. M. (2010). Is domestic violence a cultural tradition?...and other questions about gender and migration. Retrieved from http://www-rohan.sdsu.edu/~oespin/downloads/Is_domestic_violence_a_cultural_tradition.pdf

Espín, O. M. (1999). *Women crossing boundaries: A psychology of immigration and the transformations of sexuality*. New York: Routledge.

Freire, P. (1997). *Pedagogy of the oppressed*. New York: Continuum.

Guillemin, M. & Gillam, L. (2004). Ethics, reflexivity, and "ethically important moments" in research. *Qualitative Inquiry, 10*(2), pp. 261–280.

Hanisch, C. (1969). The personal is political. Retrieved from http://www.carolhanisch.org/CHwritings/PersonalisPol.pdf

Herman, J.L. (1997). *Trauma and recovery*. New York: Basic Books.

Homeland Security Office of Immigration Statistics (2011). In Yearbook of Immigration Statistics, 2012. Retrieved from http://www.dhs.gov/sites/default/files/publications/immigration-statistics/yearbook/2011/ois_yb_2011.pdf

Hondagneu-Sotelo, P. (Ed.) (2003). *Gender and U.S. Immigration: Contemporary Trends*. Berkeley, CA: University of California Press.

Hughey, J., Peterson, N.A., Lowe, J.B. & Oprescu, F. (2008). Empowerment and sense of community: Clarifying their relationship in community organizations. *Health Education & Behavior, 35*(5), pp. 651–663.

Kelly, U. (2009). I'm a mother first: The influence of mothering in the decision-making processes of battered immigrant Latino women. *Research in Nursing & Health, 32*, pp. 286–297.

Kofman, E., Phizacklea, A., Raghuram, P. & Sales, R. (2000). *Gender and international migration in Europe: employment, welfare and politics*. New York: Routledge.

Kuper, A., Reeves, J.S. & Levinson, W. (2008). An introduction to reading and appraising qualitative research. *British Medical Journal, 337*(288), pp. 404–409.

Macias, R. L. (2013). *Adaptation of a leadership training program for latinas* (Master's thesis). Retrieved from http://digitalarchive.gsu.edu/psych_theses/105

Mahler, S.J. & Pessar, P.R. (2006). Gender matters: Ethnographers bring gender from the periphery toward the core of migration studies. *International Migration Review, 40*(1), pp. 27–63.

Mankowski, E.S., Galvez, G. & Glass, N. (2011). Interdisciplinary linkage of community psychology and cross-cultural psychology: History, values, and an illustrative research and action project on intimate partner violence. *American Journal of Community Psychology, 47*(1–2), pp. 127–143.

Martín-Baró, I. (1996). *Writings for a liberation psychology*. Cambridge, MA.: Harvard University Press.

Martinez, C.R., McClure, H.H., Eddy, J. & Wilson, D. (2011). Time in U.S. residency and the social, behavioral, and emotional adjustment of Latino immigrant families. *Hispanic Journal of Behavioral Sciences, 33*(3), pp. 323–349.

Moracco, K.E., Hilton, A., Hodges, K.G. & Frasier, P.Y. (2005). Knowledge and attitudes about intimate partner violence among immigrant Latinos in rural North Carolina. *Violence Against Women, 11*(3), pp. 337–352.

Moustakas, C. (Ed.) (1994). *Phenomenological research methods*. California: Sage Publications.

NVivo qualitative data analysis software (Version 9) [Computer software]. QSR International.

Odem, M.E. & Lacy, E. (2009). *Latino immigrants and the transformation of the U.S. South*. Athens, GA: University of Georgia Press.

Perilla, J.L. (1999). Domestic violence as a human rights issue: The case of immigrant Latinos. *Hispanic Journal of Behavioral Sciences, 21*, pp. 107–133

Perilla, J.L., Serrata, J., Weinberg, J. & Lippy, C.A. (2012). Integrating women's voices and theory: A comprehensive domestic violence intervention for Latinas. *Women & Therapy, 35*(1–2), pp. 93–105.

Pessar, P.R. & Mahler, S.J. (2003). Transnational migration: Bringing gender in. *International Migration Review, 37*(3), pp. 812–846.

Portes, A. & Rumbaut, R.G. (2006). *Immigrant America: a portrait*. California: University of California Press.

Prilleltensky, I. (2012). Wellness as fairness. *American Journal of Community Psychology, 49*(1–2), pp. 1–21.

Rappaport, J. (1987). Terms of empowerment/exemplars of prevention: Toward a theory for community psychology. *American Journal of Community Psychology, 15*(2), pp. 121–148.

Riemen, D. J. (1986). The essential structure of a caring interaction: Doing phenomenology. In P.M. Minhall & C.J. Oiler, (Eds.), *Nursing research: A qualitative perspective* (pp. 85–108). Norwalk, CT: Appleton-Century-Crofts.

Rodriguez, R., Serrata, J., Rosales, A., Perilla, J. & Macias, R.L. (2012). *Exploring the intersection of immigration and domestic violence in Latino families: A participatory action research approach*. Manuscript in preparation.

Serrata, J. (2012). *Creating an opportunity for self-empowerment of immigrant Latina survivors of domestic violence* (Doctoral dissertation). Retrieved from http://digitalarchive.gsu.edu/cgi/viewcontent.cgi?article=1102&context=psych_diss

Speer, P.W., Jackson, C.B. & Peterson, N.A. (2001). The relationship between social cohesion and empowerment: Support and new implications for theory. *Health Education & Behavior, 28*(6), pp. 716–732.

Taylor, P., Lopez, M.H., Passel, J. & Motel, S. (2011). *Unauthorized immigrants: Length of residency, patterns of parenthood*. Washington, DC: Pew Research Center, Hispanic Research Center.

Tjaden, P. & Thoennes, N. (2000). *Full report of the prevalence, incidence, and consequences of intimate partner violence against women: Final report from the*

National Violence Against Women Survey. Washington, DC: U.S. Department of Justice.

Vidales, G.T. (2010). Arrested justice: The multifaceted plight of immigrant Latinas who faced domestic violence. *Journal of Family Violence, 25*(6), pp. 533–544. doi:10.1007/s10896-010-9309-5.

Worell, J. & Remer, P. (2003). *Feminist perspectives in therapy: Empowering diverse women* (2nd ed.). New York: John Wiley & Sons, Inc.

Zarza, M.J. & Adler, R.G. (2008). Latina immigrant victims of interpersonal violence in New Jersey: A needs assessment study. *Journal of Aggression, Maltreatment, & Trauma, 16*(1), pp. 22–39.

Zimmerman, M.A. (1995). Psychological empowerment: Issues and illustrations. *American Journal of Community Psychology, 23*(5), pp. 581–599.

Part IV

Intergenerational Impact of Migration

Introduction to Part IV

Oliva M. Espín and Andrea L. Dottolo

In recent decades, a significant line of psychological research has resulted from the work of a few psychologists on the importance of generation, specifically on the intersection of sociohistorical context and life stage. Similar to Espín (1997; 1999), Stewart (2003) underscores the idea that "Thinking about generation allows us to see how our own personal history is aligned with social history" (p. 3).

Stewart and Healy (1989) developed a model for studying the relationship between individual and social histories in a unique amalgamation of theories by Mannheim and Erikson. Mannheim (1952) defined a generation as a group of people experiencing similar social events in a similar time and place and of the same age. The social events that are particularly relevant in defining a generation happen when the primary cohort is between the ages of around 17 and 25 – that is, in the critical period – according to Mannheim (1952). Stewart and Healy (1989) extended Mannheim's theory to incorporate other life stages or periods of development, "as Erikson (1968) did," stressing that it was *intersection* of life stage and social events that shaped personality. Stewart and Healy (1989) argued that events that occurred during early childhood would affect fundamental expectations and values. For example, individuals who were children during migration and experienced familial disruptions might translate these early experiences into what they might expect from their environments later in life. For those who were in late adolescence or early adulthood, these events affect identity formation and perceptions of opportunities. Events experienced during early to middle adulthood might influence behavior, but according to the theory, because work and family obligations may take precedence at this life stage, the events will be less likely to affect personality. Lastly, events that happened during midlife were hypothesized to spark alterations or reconsiderations of identity for some individuals, as they may witness new choices and opportunities potentially available to them later in life with the immediacy of career and/or family duties at bay.

Stewart and Healy (1989) qualify this hypothesis by suggesting that life events such as migration can affect different people in various ways. This model is especially useful in examining the psychological impact of migration on generations or cohorts. In addition, it seems useful to us as a backdrop for understanding the impact of the experience of migration on US-born descendants of immigrants. For the most part, if migration is considered at all in psychology, the prevalent idea is that immigration may affect the immigrant, but descendants of immigrants who are already citizens are not considered to be affected by this process at all. Contrary to that perspective, we believe that the effects of migration are felt through several generations, which is why we are including a section on this topic in this book.

An important series of studies have applied Stewart and Healy's (1989) innovative model to different groups and events. For example, Duncan and Agronick (1995) found that social events played a primary identity-shaping role when experienced in early adulthood. In particular, the women's movement was especially relevant for women who were young adults at its height, as well as women in midlife who became rejuvenated by its momentum and significance in their lives. Similarly, Duncan and Stewart (2000) demonstrated how three generations of feminists shared similar political commitments with varying identity formations based on cohort. Hurtado, Gurin & Peng (1994) showed the influence of generation on identity structure, where Chicanos (second generation or later) had more differentiated identities than Mexicanos (first generation immigrants). Cole & Stewart (1996) examined the ways in which the women's movement and the civil rights movement affected both black and white women.

Sociologist Richard Jenkins illustrates an important psychological component of social identities as located primarily in the minds of individuals, claiming that "Although…identities are imagined; they are not imaginary" (1996, p. 175). Characterizing identities as "imagined" depicts them as social constructions, though indelibly "real". In most discussions of contemporary social identity theory, scholars often credit much of the groundwork to psychologist Henri Tajfel, who describes social identity as "that part of an individual's self-concept which derives from his knowledge of his membership of a social group (or groups)" (Tajfel, 1981).

Histories of US immigration have created tension around racial categorizations. In her discussion of "how the Jews became white folks,"

Brodkin (2004) explains, "The postwar period was a historic moment for real class mobility and for the affluence we have erroneously come to believe was the American norm. It was a time when the old white and the newly white masses became middle class" (p. 23). Similarly, Jacobson (1998) explains three shifts or developments in racial politics after World War II that altered racial and ethnic politics and ideologies. His first assertion is that "ethnicity" was adopted to signify cultural difference and helps to explain why those white groups who had been in the United States the longest, such as the Irish, Dutch, English and Germans, could claim an ethnic/cultural identity without ever compromising their whiteness. This is extended into his second assertion, that race, "deemed irrelevant," now applied to "others" and enabled the construction of discourses of "colorblindness". Third, consolidating whiteness also contributed to the consolidation of blackness, "giving way to a strict scheme of black and white," planting the seeds for the civil rights movement (p. 110).

Focusing on migration, the connection between age and the impact of this life event is self-evident. But how does this process affect their descendants? For daughters of immigrants, it may become difficult to view their mothers in a positive light because they do not seem to embody the traits and characteristics that an "American" woman should. In many cases, migration has limited the mother's sphere of influence even further (Espín, 1999; 2006). Frequently, women are made to "carry" cultural values and behaviors at the expense of their own lives. Cultural pressures to be the "bearers of culture" may bring significant stress to the lives of women. Virtues such as dedication to family and husband, for example, place a great burden on married women's shoulders. Family and cultural celebrations translate into long hours of cooking and house tending. Daughters are expected to help mothers in these tasks, further constraining their activities and available time. Sons are frequently free to go and enjoy celebrations without the expectation that they will help with the tasks at hand.

Also, the development of the daughters' identity as young women becomes intertwined with the acculturation process that the parents are undergoing and with other emotional demands placed on them. For daughters of immigrants, the process includes the understanding and adoption of more than one cultural context and identity. "If sex roles are problematic for adolescent females in general, this is doubly true for girls in immigrant families. They are faced with having to adjust to two sex cultures" (Goodenow & Espín, 1993). As they construct their cultural identity as American adolescents, they simultaneously develop a sexual identity. Emotional resources are strained by the need to cope with two demanding life tasks of primary importance for adolescents: psychological development into healthy adulthood and the formation of identity.

The chapters in this section address ways in which the effects of immigration are present and linger through the generations, even for those who have not experienced the process of migration themselves.

The problematic situation of children of undocumented immigrants, some of whom immigrated with their parents at a very young age and some of whom were born in the US who therefore are US citizens, is very actively present in the national conversation. The unauthorized migrant population, often referred to as *"illegal"* or *"undocumented"* immigrants, represents individuals who seek to relocate to the US in search of employment and better living conditions. This population includes those entering the US without inspection and those entering the US with legal temporary visas but staying beyond the time allotment of their visas. It is estimated that 10 to 12 million individuals reside in the US without legal documents. While the US prides itself on being "a nation of immigrants", reception of those seeking a new home in this country has not always been as welcoming as some idealized accounts purport it to be.

The chapter by Stutman and Brady-Amoon focuses on the US citizen daughters of Ecuadorian immigrant to the US. The authors explore the lived experiences and stressors of 39 first-generation Ecuadorian American girls as told through the narratives that accompany their house-tree-person (HTP) drawings. These evaluations were done at a moment in time when one or both parents has emerged from hiding in order to attempt to gain legal residential status in the United States. This chapter illustrates one of the issues in the intergenerational transmission of immigration stresses.

Sandra Mattar and María Teresa Pestana, a mother-daughter duo, address some aspects of the gendered immigration and acculturation gap in the literature by discussing the need to understand migration narratives across generations and the impact of multiple contextual factors on acculturation processes. Using a qualitative case study method, the chapter describes the immigration narratives of two generations of women impacted by voluntary transnational serial migration. The authors address the need for more clarity on the role of social class, sexual orientation and language use, among others, in determining the outcomes of immigration experiences in order to avoid essentialist immigration narratives. They recommend the use of an intersectional analysis to inform the field of gendered immigration studies rather than the very linear and simplistic terms in which acculturation is described in the research literature, the latter of which ignores the multiple pathways that cultural identity can take as well as the gendered, racialized and politicized experiences of immigrants.

Karen Kisiel Dion considers the challenges that gendered expectations can create for daughters from immigrant families as they construct a sense of their own identity. Based on interviews with university undergraduate women born in Canada whose parents immigrated to that country from several different societies, this chapter portrays a perspective on self-definition in daughters of immigrants that contributes to an understanding of gender and cultural identity. This is the only chapter in the book that discusses immigration-related experiences outside of the United States. However, we believe that despite differences between immigration policies between Canada and the US, the experience of citizen daughters of immigrants with respect to the psychological impact of migration on the development of gender and cultural identity is similar enough to merit inclusion in this collection.

Andrea L. Dottolo and Carol Dottolo present another noteworthy daughter-mother collaboration. Their study focuses on how granddaughters of immigrants construct and maintain their ethnic identity – in this case, being Italian. Since language cannot serve as a symbol for these women, because speaking Italian was often forbidden in their homes or spoken only between adults in covert communications, they must cling to other symbols of Italianness in order to preserve their sense of gendered ethnic identities. Dottolo and Dottolo argue that one such symbol is food, wherein participants manipulate recipes and food to navigate and negotiate their Italian and American identities, femininity, social class and honoring of both a lost and imagined cultural experience.

Together, the chapters in this last section of the book offer a point of view rarely acknowledged in studies of migration, that of descendants of immigrants whose lives are impacted by their ancestors' decision to migrate to another country. Some chapters refer to the direct experience of daughters of immigrants, where the effects of parental migration is obviously more immediate. But, the chapter by Dottolo and Dottolo shows that the psychological impact of migration continues to be present after several generations.

References

Brodkin, K. (2004). How did Jews become white folks? In M. Fine, L. Weis, L.P. Pruitt & A. Burns (Eds.), *Off white: Readings on power, privilege, and resistance* (pp. 17–34). New York: Routledge.

Cole, E.R. & Stewart, A.J. (1996). Meanings of political participation among Black and White Women: Political identity and social responsibility. *Journal of Personality and Social Psychology, 71*, pp. 130–140.

Duncan, L.E. & Agronick, G.S. (1995). The intersection of life stage and social events: Personality and life outcomes. *Journal of Personality and Social Psychology, 69*(3), pp. 558–568.

Duncan, L.E. & Stewart, A.J. (2000). A generational analysis of women's rights activists. *Psychology of Women Quarterly, 24*, pp. 297–308.

Erikson, E.H. (1968). *Identity, youth and crisis.* New York: W.W. Norton & Company.

Espín, O.M. (1997). *Latina realities: Essays on healing, migration, and sexuality.* Boulder, CO: Westview Press.

Espín, O.M. (1999). *Women Crossing Boundaries: A Psychology of Immigration and the Transformations of Sexuality*. New York: Routledge.

Espín, O.M. (2006). Gender, sexuality, language and migration. In R. Mahalingam (Ed.), *Cultural psychology of immigrants* (pp. 241–258). Mahwah, NJ: Lawrence Erlbaum.

Goodenow, C. & Espín. O.M. (1993). Identity choices in immigrant female adolescents. *Adolescence, 28*, pp. 173–184.

Hurtado, A., Gurin, P. & Peng, T. (1994). Social identities: A framework for studying the adaptations of immigrants and ethnics: the adaptations of Mexicans in the United States. *Social Problems, 41*, pp. 129–151.

Jacobson, M.F. (1998). *Whiteness of a different color: European immigrants and the alchemy of race.* Cambridge, MA: Harvard University Press.

Jenkins, R. (1996). *Social identity*. New York: Routledge.

Mannheim, K. (1952). The problem of generations. In *Essays on the sociology of knowledge.* New York: Oxford University Press. (Original work published in 1928).

Stewart, A.J. (2003). 2002 Carolyn Sherif Award address: Gender, race, and generation in a Midwest high school: Using ethnographically informed methods in psychology. *Psychology of Women Quarterly, 27*, pp. 1–11.

Stewart, A.J. & Healy, J.M., Jr (1989). Linking individual development and social change. *American Psychologist, 44*, pp. 30–42.

Tajfel, H. (1981). *Human groups and social categories: Studies in social psychology.* Cambridge: Cambridge University Press.

10

Assessing and Redressing Effects of Second-Class Citizenship upon US Citizen Daughters of Undocumented Immigrants

Gabrielle Stutman and Peggy Brady-Amoon

This chapter focuses on the US citizen daughters of undocumented Ecuadorian immigrants to the US. The information reported comes from a moment in time when one or both parents of these US citizen children has emerged from hiding in order to attempt to gain legal residential status in the United States. As described in our earlier work (Stutman & Brady-Amoon, 2011), a psychological hardship evaluation indicating their citizen children would experience *extreme and unusual psychological hardship* if this attempt is unsuccessful may support an *application for change of status* for one or both parents that keeps the family together. This is a time of great stress because it requires that the family risk deportation as these legal proceedings may turn against them.

It was through my assessments of over 200 families that I, Gabrielle, had the opportunity to meet and get to know many of these citizen daughters. Consequently, I became increasingly aware of the trauma that living with the threat of deportation has had upon these children. The information presented about this population is a distillation of my work with 76 of these families between 2007 and 2011. That work included evaluations of the moral, cognitive and psychological health of Ecuadorian parents and psychological assessments of 39 five- to fourteen-year-old, first-generation, US citizen Ecuadorian American girls.

The process of this evaluation is fully described in our earlier work (Stutman & Brady-Amoon, 2011) and briefly reviewed in this chapter. It is informed by a feminist analysis, which addresses the understanding of power, intersecting identities in context and a commitment to the eradication of oppression, particularly of women and girls (Espin, 1999; FTI, 1999). We are also informed by an integrative humanistic perspective, which brings together feminist, humanist, multicultural and social justice principles and practices (Brady-Amoon, 2011). These perspectives respect the intersection and influence of gender, race, language, age and acculturative status on people's lived experiences (Espin, 1999; Hondagneu-Sotelo, 2003; Pessar, 2003; Thorne, Orellana, Lam & Chee, 2003). They also inform psychological practice, including assessment and evaluation. Consequently, we strive to respect people's narratives and behavior within the context of their lived experiences. In keeping with a feminist and integrated humanistic perspective, we work to move beyond the limitations of the medical model, eschewing diagnosis when practicable, recognizing that symptomology is often an adaptive response to an unhealthy environment or to unhealthy systems, particularly for people who have suffered trauma and marginalization (Caplan, 1995; Chesler, 1972/2005; Kirschner, 2013; Morrow, Hawxhurst, Montes de Vegas, Abouselman & Castaneda, 2006; Rogers, 1980).

At the same time, our experience as psychologists working in forensic settings has shown that we are most effective when we infuse feminist principles, including consideration of our own identities and social locations, with traditional practice guidelines, including the American Psychological Association's (APA) guidelines for psychological practice with girls and women (2007) and the APA specialty guidelines for forensic psychology (2013). This considered approach, integrating both quantitative and qualitative methodology (Crawford, 2013), has proven to be successful in influencing the US immigration system to support family unification (Stutman & Brady-Amoon, 2011).

In the sections that follow, we offer a brief review of immigration to the US and some of the reasons for the Ecuadorian diaspora. Next, we offer an examination of Ecuadorian culture as related to child-rearing practices, which will set the stage for an exploration of the stressors of first-generation Ecuadorian American girls as they become aware of, and attempt to negotiate the disjunction between their culture of origin, their parents' expectations and the unique danger they face in the context of their parents' decision to seek a *change of status*. We conclude with recommendations for practice and policy.

Immigration and the United States of America

The United States is a nation of immigrants. Approximately 37.6 million people, comprising 12.3 percent of the US population, are immigrants

(U.S. Census, 2010). Children under 18 years of age comprise about 8.9 percent or 2.4 million of these immigrants (U.S. Census, 2003). Furthermore, nearly a quarter of children under 18 in the US have an immigrant parent. Today, more than half the immigrant population comes from Latin America. Although more than 1.8 million refugees and asylum seekers were admitted to the US in the last 25 years, the UN reports that there are 9.9 million unplaced people who meet the criteria for these classifications (Short, Suzuki, Prednes-Lintel, Furr, Madabbushi & Mapel, 2010).

New immigrants come to the US for the same reasons that most of our immigrant ancestors came. They come to escape wars, persecution, extreme poverty and other injustices, and they come to seek employment and educational opportunities (Short et al., 2010). Advances in technology and transportation have changed the experience of immigration such that today's immigrants are less likely to sever all communication with family and friends in the "old country". To some degree, this is a protective phenomenon. Nevertheless, trauma and loss are not uncommon before, during and after immigration, even for people not formally classified as refugees (Espin, 1999). This may be particularly true for women and girls whose experiences are often unrecognized, unsupported and devalued (Espin, 1987; 1997).

For many immigrant families, intergenerational differences are compounded as children master English and acculturate more quickly to their new environment than their elders do (APA, 2012; Rivera, 2010). However, children's adaptation and adjustment is often more complex than generally believed (Espin, 1999). Researchers are just beginning to consider the experiences of the children of immigrants (Thorne et al., 2003), including the effects of US immigration policy on the citizen children of undocumented immigrants (Capps, Castaneda, Chaudry & Santos, 2007; Chaudry, Capps, Pedroza, Casteneda, Santos & Scott, 2010).

We have previously described the negative effects of current US law and policy, which include wholesale arrests, detentions and deportations. These may follow such minor infractions as a traffic violation or any other brush with the law that reveals their undocumented status. When this happens, their children are faced with a situation in which they will either lose a parent or be transported along with their parent(s) back to the country of origin (Stutman & Brady-Amoon, 2011). This is not an unrealistic fear as approximately 200,000 parents of children who are US citizens were deported between 2010 and 2012, leaving more than 5,100 US citizen children in foster care (Garcia, 2013). For those who stay with a remaining parent, abandonment anxiety focused on losing that parent sharply escalates. We turn now to the context of emigration from Ecuador today.

Emigration from Ecuador: Ecuadorian women and their citizen daughters

Ecuadorian politics and society have long been characterized by conflict, disruption and instability (Human Rights, 2012; Political Conditions, 2012). More than 40 percent of the Ecuadorian population lives below the poverty line (Human Rights, 2012), illustrated by the fact that 26 percent of children under age five are diagnosed with stunted growth (Larrea & Kawachi, 2005). Furthermore, the lack of clean water and sanitation facilitates the spread of disease (Harari & Harari, 2006). A corrupt police force and judiciary; widespread systematic discrimination against indigenous people, LGBT persons and the poor (Human Rights, 2012); and an influx of people involved with and fleeing from violent conflict in neighboring Colombia has resulted in widespread emigration from Ecuador (Orces, 2009). Therefore, although not formally recognized as such, many present-day Ecuadorian immigrants are refugees (Espin, 1999).

For the most part, immigrants who lack official authorization have low socioeconomic status and poor English fluency. Consequently, their capacity to negotiate the complexities of immigration law is compromised. Those Ecuadorian immigrants interviewed in their quest for a *change of status* were among the more successful. All had lived in the US for ten or more years, most having arrived when they were in their 1920s or early 1930s. They had sufficient resources to hire an attorney, pay government fees and pay for the services of an independent psychologist. Our practice is to conduct interviews in the language preferred by the client. As such, most adult interviews and all the children's interviews were conducted in English. Many adult clients chose to do the mental status examination in Spanish to maximize their performance. Adults who preferred to be interviewed in Spanish were interviewed using a qualified interpreter as Gabrielle's Spanish language skill is less than fluent.

Many of the Ecuadorian immigrant adults interviewed described how gangs flourish in the cities of Ecuador and how violence is rampant even in the Ecuadorian countryside. They report that public buses are routinely boarded by knife or gun-toting thieves who are usually able to divest the passengers of their valuables without interference. Private cars and taxis are stopped, the occupants removed and sometimes kidnapped at gunpoint. Sadly, even shoes and clothing are stolen, leaving the victim naked on the road. No one in Ecuador steps outside the house wearing jewelry. On the rare occasion when a perpetrator is caught, the police are often ready to accept a bribe to release the felon.

The disjunction between these events and the basically peaceful nature of the Ecuadorian people is best exemplified by the way in which they raise their children. Ecuadorian parenting is characterized by positive reinforcement, which includes induction, monitoring and encouragement (Ingoldsby, Schvaneveldt, Supple & Bush, 2004). Clear boundaries and expectations are modeled, and very few Ecuadorian parents use corporal punishment (Rae-Espinoza, 2010). Parents who were interviewed generally described their response to inappropriate child behavior as withholding privileges. Some merely "talk with" their children, and a few parents of older children could not remember the last time that corrective action was necessary. Weekends are generally spent with the family unit, even when the children are of high school age. This child-rearing pattern is generally believed to foster secure emotional attachment and clear moral/ethical judgments (Ingoldsby et al., 2004; Rae-Espinoza, 2010). Indeed, the children interviewed manifested an unusually strong attachment to their parents and early social consciousness.

However, this close family unit, and even their exemplary child-rearing patterns are often unable to compete with the current devastation of Ecuadorian society. Clients report that the lack of productive employment makes joining a gang very attractive to adolescents in Ecuador. Children are generally out of school by the age of 12, and crime is often the only way to earn money for essential needs. Those who would not join willingly are often coerced into membership. Boys are forced to join by being beaten and/or in response to threats to their family members, which has also been reported in the press (Kraul, 2011). Girls are reported to be sexually harassed. They may be bribed with food or clothing, but also they may be gang-raped as an inducement to join and/or as an initiation rite. Once initiated, they will become the property of a male gang member who will act as their "protector". Girls as young as ten years old are reported to have become pregnant in this way. Once enlisted, a pattern of drug use and crime becomes a way of life.

These factors have combined to motivate a constant stream of Ecuadorian migration. Some of the parents interviewed arrived with tourist visas and simply did not return home. Most paid an average of $6,000 per person and risked their lives to be shepherded across the Mexican border by "coyotes". They arrive with little to no English language skills and a lack of formal education relative to US standards.

This is particularly true for women and girls. In Ecuador, school books and uniforms must be provided by the child's parents from the first grade, and tuition must also be paid after the sixth grade. Most children are not educated beyond that point. When resources are scarce, boys in Ecuador tend to be given educational priority because they are seen as the potential providers for their family.

In our sample of 76 Ecuadorian immigrant families in which one or both parents were undocumented, the mean educational level was nine years for adults of both sexes. However, four of the women (but none of the men) had less than a sixth grade education. One woman, who had attended only one year of school opined, "That's the way it is in the mountains." The most educated woman had 13 years of school (one year of college), while three of the men exceeded that number of years of schooling. One man had completed college but still couldn't find a job in Ecuador.

This formal educational deficiency and lack of employment opportunity compromises the prosperity of both men and women, but it has a predictably greater negative effect upon the women, even after their arrival in the US. Although the men interviewed usually started as unskilled factory or construction workers, over time they were more likely to develop English language and working skills. The women were likely to remain in less remunerative work, such as house or office cleaning. They were less likely to progress in English or in the workplace, and they have tended to become full-time homemakers after the birth of their first child. One positive effect of this disjunction is that the women are generally a constant presence in the household. This helps to create and maintain the strong bonds that secure their family life.

Although the men tend to work long hours outside the home during the week, time off is spent as a family. Sundays generally begin with a family breakfast, followed by attendance at church and family outings to a park, beach, or the mall for shopping. If extended family members live nearby, the adults visit and the cousins play together. Because they lack documents that identify them to the larger society and are fearful of discovery and deportation, this strong bond is essential to the emotional security of the family.

Also of great importance is their clear expectation of their children's success. This is not only true for the boys. Girls are not groomed to follow in their mother's footsteps. Virtually every interviewed girl over the age of ten foresaw a college education leading to a professional career of some kind. They are a highly motivated cohort, worried about tests and grades and deeply engaged in justifying their parents' sacrifices for them. When asked what they would do if given a "big barrel of money", the frequent response was to buy a big house for their parents, pay for college and give money to charity.

Despite their closely knit "overprotective" family unit, undocumented parents cannot help but communicate anxiety and helplessness to their vulnerable offspring. They assiduously attempt to protect their daughters from these threats. The younger girls are unconscious of the source of their parents' anxiety, but they absorb the emotional tone of the household, and a free-floating anxiety is often the likely result. By the time they reach nine or ten years of age, they have become acutely aware of their parents' secret, their lack of safety, their marginalized identity and the need to remain hidden. As they continue to mature, they become even more aware of how their parents differ from those of their peers. Awareness of second-class status may derive from a situation as seemingly innocuous as the inability to get a library card or as serious as fear of reporting a crime to the police. As they mature, they understand more deeply the need to hide from the authorities. They hear about a friend's parent or a relative who has been deported and learn to fear that they also might never see a detained parent again.

In addition to separation anxiety, these girls are economically at risk when a parent, usually the father, is deported. The mother's presence as a full time homemaker makes her financially dependent. If a male head of household is deported, she has little choice but to accompany him. She cannot survive here alone with their children, and so the children must either leave their country or lose their parents.

Documenting extreme hardship: Girls' drawings and their stories

In this section, we explore the responses of 39 five- to fourteen-year-old US citizen daughters of undocumented Ecuadorian immigrant parents to the house-tree-person (HTP) protocol and their narratives. The HTP and accompanying narratives contribute to a comprehensive psychological hardship evaluation that may be used to document a citizen child's *extreme and unusual hardship* should a parent's application for a *change of status* be denied.

The HTP is the most popular projective assessment used by psychologists (Safran, 1997) and is widely used in schools (Esquivel, Oades-Sese & Olitzky, 2008). Projective assessments are based on the theory that people's responses to ambiguous stimuli offer insights into their personality and unconscious motivations. Building on Goodenough's (1928) draw a man/draw a person assessment, originally considered a test of intellectual functioning, Buck (1946; 1969) developed the HTP, theorizing that people's drawings – of houses, trees and persons – were insights into their personality and psychological functioning

The HTP can be administered to anyone over age three who can draw. The only tools that are necessary are clean white paper and a pencil. The instructions are also simple. The administrator simply asks the participant to draw a house and, when that is done, to draw a tree, and finally, to draw a person (WPS, 2013).

There are multiple ways to score and interpret HTP drawings. Buck (1946) and Koppitz (1968) developed and manualized psychodynamic interpretations. There are also quantitative/objective scoring protocols that suggest that normally developing and healthy functioning people are expected to, and almost always will, provide the "essential details" of the subject they are drawing (Ogdon, 1981). For example, although not culturally or geographically universal, the essential details of a house are traditionally described as a wall, a roof, a window, a door and a chimney. Regarding psychoanalytic interpretations, knotholes in trees are believed to suggest trauma because the imagined tree appears injured (Buck, 1946; Devore & Fryrear, 1976). However, there are other plausible interpretations. "Holistic-intuitive" approaches seek to integrate psychoanalytic and objective interpretations (Safran, 1997, p. 3).

Given our goals and orientation described earlier, our feminist, multicultural approach integrates the above with narrative approaches in which the artist is invited to describe her drawings in dialogue with the examiner (Esquivel et al., 2008). We have found that most children enjoy drawing and talking about their drawings. At the same time, sometimes it helps to ask open-ended questions about the drawings, some of which are suggested in scoring manuals.

Because "no knowledge is acultural" (Padilla & Borsato, 2008, p. 5), we use a feminist, ecological and culturally informed perspective to generate our questions and conclusions. In so doing, we strive to attend to cultural, social and gendered influences on drawings as well as our own interpretations – recognizing, for example, that houses in warmer climates tend not to have chimneys and so will rarely be drawn with one. Furthermore, because these interviews, described earlier and in our previous work (Stutman & Brady-Amoon, 2011), were designed for submission to the court as part of the applicant's application for *change of status*, the process is informed by the APA (2013) forensic guidelines.

In so doing, we are aligned with Crawford's (2013) position that all research – and we would contend, all evaluation methods – are potentially feminist. Specifically, we have found that quantitative methodology combined with a narrative, qualitative approach has proven most effective in our forensic setting. To illustrate, the court traditionally places weight on a child's physical or mental disability to assess their hardship as not only extreme, but also unusual, and thus worthy of a waiver for their parents. Our use of multiple methods, as further described below, has shown that children who are mentally very well endowed are likely to experience the threat of relocation earlier, with greater sensitivity and with more emotionally psychopathological effect. Therefore, their hardship might also be judged as extreme and unusual.

Procedure

Psychologists engaged in forensic work – that is, work designed to inform court officers or proceedings – are expected to maintain a neutral expert stance (APA, 2013). As such, before beginning the forensic evaluation, I, Gabrielle, routinely engage in an informed consent process. I explain my role as a neutral evaluator and the importance of including negative as well as positive disclosures, because even apparently negative information, such as a prior arrest, will enhance the utility of the report and thus aid their case.

Much feminist literature focuses on consciousness of and leveling the traditional hierarchy between professional (e.g., researcher) and participant, in the service of empowering women's voices by hearing them on their own terms (Yost & Chmielewski, 2012). As such, I also acknowledge and welcome exploration of power differentials in this evaluation process, including my status as a formally educated psychologist, doctor and US citizen and as an older, white, Jewish, female "outsider" (Wilkinson & Kitzinger, 2013). I also address the clients' perceptions of my role and identities – including the likely perception that I have more social and political power than they have, particularly as undocumented immigrants seeking a *change of status* to legally reside in the US, as well as how that power is likely to be helpful to them and their children.

With appropriate consent, an extensive interview is conducted with all members of the family who are able to be present. This includes a biopsychosocial history of the parents and their children, a mental status evaluation of the parents and a review of what the family situation would likely be if one or more parents were to be deported. Because the children remain present throughout most of the initial assessment, I am able to form a relationship with them prior to their time alone with me.

The child assessment includes a child interview, a sentence completion task, and the HTP. I generally begin by asking the child if she knows why she is here with me and then continue the interview. The questions first elicit pleasant fantasies and later progress to more potentially anxiety-prone issues. The sentence completions are presented as a finish-the-sentence game. When drawing, they are first asked to draw the best house they can, and then they're told, "Tell me about your house." Next they are asked to draw and describe their tree and, finally, their person. In this way, a play-like atmosphere is created, at least for the younger girls, as most enjoy answering fantasy questions, playing simple word games and drawing. The older girls, who are more aware of the reason for the evaluation, are initially rather more anxious, but they generally relax as we form a relationship in the context of my positive responses to their answers and drawings.

In the section below, we report the girls' descriptions of, and responses to, open-ended questions about their drawings. As described above, the girls' narratives are presented in parallel with psychoanalytic interpretations (Buck, 1946; 1969; Koppitz, 1968) and normative scoring protocols (Ogdon, 1981). This is consistent with psychoanalytic as well as feminist, multicultural approaches. The occupants of the house, the location of the house and tree, the age of the tree and person and the activity of the person were usually obtained. These narratives, along with the children's previous verbal responses, informed the interpretations.

Children's drawings depart from adults' drawings in many ways. As one may expect, they are generally not as well developed, and the skill level is likely to be somewhat less. A normal developmental progression enables a child first to draw a tree, then a house and, finally, a person (Kalyan-Masih, 1976; Leeds, Dirlam & Brannigan, 1983). Although the sequence is universal, the rendering will be influenced by visuospatial and fine motor skill development and training. Going to school and having older siblings will tend to advance development in this as well as in most other areas of cognitive competence. Cultural and gender differences may be reflected in their narrative or the drawing itself.

Analysis

In reviewing all 39 sets of HTP drawings by using a modified qualitative categorization method, we categorized the girls' drawings as typical when the content and form was shared by at least 65 percent of the sample, general as shown by 26–64 percent of the sample, or variant when the content or form was shown by less than 25 percent of the sample.

The girls' house drawings typically diverge from previously established norms in that they were as likely to draw a large, aspirational house as a moderately sized one. This non-normative tendency to draw large houses may reflect the girls' familial role to justify their parents' sacrifice via their success. Small drawings were variant in our sample and are traditionally considered suggestive of psychopathology according to both normative and psychoanalytic approaches (Ogdon, 1981). The narrative of five year-old Jennifer, whose drawings were less than one square inch in size, supported this hypothesis. Jennifer is the youngest of four siblings and the only child born in the US. Of three siblings left behind in Ecuador, one died because her father was unable to feed the child. She exhibits significant psychopathology and is described by her parents as uninterested in her environment, shy and easily led. Furthermore, she reported her person as only three years old, which suggests strong regressive urges according to psychoanalytic and normative interpretations (Ogdon, 1981).

The chimney of the house, although often regarded as a basic elaboration that suggests familial warmth (Ogdon, 1981), was rarely drawn by these girls, even when they lived in a private house in the Northeast United States. In addition to cultural influences, this may be due to the fact that newer houses no longer have the large brick chimneys that were once the norm. A three-dimensional house, which was also variant, was never seen in a child who was less than eight years old and never in more than 23 percent of the sample. This variant appeared more meaningful as it was highly associated with children who appeared cognitively age-advanced as judged from their verbal responses.

Michelle, a 14-year-old eighth grader whose father had 12 years and mother had four years of formal education, is separated from two of her four older siblings who remain in Ecuador. Her age-advanced drawings include a three-dimensional house elaborated with foundation plantings but no chimney. She nevertheless, imagined it to be (warmly) populated by all of her siblings. In a similar manner, her tree is a "tall, strong Christmas tree that unites the family" and her person, an excellent student who plays sports, doesn't give up and helps the family. The normative assessment of Michelle's drawings when combined with her narratives enhances our understanding of the pressure for success placed upon her and illustrates the importance of family unity in this cohort. She functions well, but like so many Ecuadorian American girls, especially those with higher than average cognitive potential, she is painfully aware that it is her responsibility to justify and rectify her parents' sacrifice as well as that of her less fortunate siblings.

The addition of tree roots is considered a developmental achievement. This was true in our sample as the absence of roots after the age of nine was variant. After age nine, the girls' narratives confirmed that this variant represented insecurity regarding a possible loss of their parent, home and/or country. In a similar confirmation of traditional understanding (Ogdon, 1981), most girls up to age 12 placed the age of the tree as older than themselves, suggesting an identification of the tree with a parental (usually maternal) figure, but by age 13 all but two girls judged the tree to be their own age or younger, suggesting a developmental unconscious identification with the tree.

Cloud-topped trees are thought to suggest a tendency toward fantasy gratification as opposed to environmental engagement (Ogdon, 1981). Contrary to expected norms, this was present in about half of these drawings, regardless of age. This suggests an unusually high rate of fantasy gratification for this cohort, and it illustrates an effect of the constant strain under which these citizen children live, now exacerbated by the current danger inherent in the pending legal procedure. Sylvia, a 12-year-old ninth grader, whose father completed 14 years and whose mother completed 12 years of formal schooling, has won numerous awards for citizenship and academic pursuits. However, her historically good grades (A's and B's) dropped significantly (C's and D's) during her father's *change of status* proceedings. Her tree is placed on the left edge of the page, suggesting regression; has a cut-off branch, suggesting trauma and hostility; a cloud-like top, suggesting fantasy; and deeply marked lines for bark on the trunk, also suggesting hostility. Sylvia described the crown of the tree as a place for birds and squirrels to hide. She wished she were five years old and constantly worries about the potential loss of her father.

As discussed earlier, a hole in the trunk of the tree is usually interpreted as representing trauma (Buck, 1946; 1969; Koppitz, 1968), and in this cohort holes were present in about half of the tree trunks. However, the associated stories almost inevitably concerned a well-stocked living space for a squirrel or baby birds. These responses suggest the girls' attempt to defend themselves against loss by generating a reparative, positive meaning of maternal nurturance for this detail.

Not so for Gladys, a 14-year-old ninth grader, whose father had six years and whose mother had one year of formal education. She drew a three-dimensional house, suggesting high average cognitive potential, and a tree on top of a mountain, suggesting high levels of aspiration. However, she uniquely described the hole in her tree trunk as a place where squirrels hide, thus revealing her fear that her parents would be discovered and deported. Consistent with her narrative, regressive longings were represented by the apples in the crown of her tree. Most poignantly, her ("happy") person is only ten years old and has no pupils in her eyes, which blankly stare out from the page. Such "empty eyes" suggest a neurotic or schizoid withdrawal from the environment (Ogdon, 1981).

As would be expected, the greatest variability in structure and content was observed in the person drawings. Nevertheless, as traditionally expected, large person drawings in this cohort were variant regardless of age and small person drawings were typical before age eight, with moderate size universal by age 13. Moderate line intensity after age eight and left-sided placement, suggesting regressive urges, were variant in girls over age nine. However, all of these girls exhibited a general tendency to place the person low on the page, suggesting low self-esteem. The use of dark/intense lines, suggesting hostility, were evident in 26 percent of drawings by girls between 11 and 14 years. Although only one 11-year-old girl drew her person with a light line, often construed as indicative of timidity, fear and inhibition, by 13–14 years old, 31 percent of girls drew their person very lightly. This supports the previous observation of a greater-than-expected and developmentally correlated diminished self-worth, inhibition and insecurity about self-expression. Although boys' drawings are expected to be more heavily rendered (Ogden, 1981), the tendency for lightly drawn representations is true of the boys in our sample as well as the girls, and it suggests diminished self-worth in the entire cohort.

As predicted, the person's head was almost universally drawn first but was often missing essential parts. Ears were hardly ever drawn, hidden as they were by the girls' long dark hair that is a cultural norm in this population. More interestingly, noses were generally missing, and lack of a nose did not become a variant until age 11. This may be a US cultural response to the popular "happy face" (☺) that also lacks a nose. Other rare variants were lack of a mouth (inhibited self-expression), drawn by a six year old and a 13 year old, and lack of pupils (withdrawal), drawn by three girls over age 11. The presence of teeth (hostility), although never more than a variant, became more frequent as the girls matured. The addition of hands (environmental competence) and feet (autonomous movement) showed a predictable developmental course, with hands becoming present a bit before the addition of feet. By age 11 a person lacking hands was variant, and by age 13 a drawing where the hands were hidden or missing was unique.

A particularly disturbing variant was the total absence of a body in the person drawings of four girls aged 9–14. These girls drew only the head (and sometimes shoulders) of their person. Wendy was a 13-year-old eighth grader whose mother had ten years of formal education. In addition to the usual stressors, her parents had divorced when she was an infant, and two of her five siblings reside in Ecuador. Although her father supports her and sees her every day, the traumatic effects of familial fragmentation in the context of deportation fears are illuminated by combining the normative and narrative approach. Her three-dimensional house (high average intelligence) with smoking chimney (familial warmth) is an aspirational three-family house where her father and (half) sister (from her father's new relationship) live with her. The other two families in the house may be understood as the two distant Ecuadorian siblings. She reports that her beautifully drawn tree can be used to make a tree house that you can climb in, but it shakes a lot (is not a stable residence). Wendy's person, described as a "good girl who smiles a lot and is all perfect", is drawn without a body. The fact that no person drawing done by a boy lacked a body suggests some girls' introjection of aspirational demands may lead them to identify with a disembodied intellect. This may be catalyzed by the larger culture, which still tends to equate masculine gender/sexuality with intelligence and success.

It was typical for girls over age eight to describe their person as actively engaged in some way and generally (aspirationally) older than themselves. As predicted, their person was generally of the same gender. However, drawing an opposite sex person, while variant and suggestive of trauma or gender dysphoria in traditional interpretations, supported an assessment of trauma but not necessarily gender dysphoria. To illustrate, Donna, an age-advanced seven-year-old second grader, whose father had six years and whose mother had nine years of formal education, has a brother in Ecuador with whom she speaks on the phone. Her person drawing, which is of her brother, reflects the emotional pain of her fragmented family. She has nightmares in which her mother and dog are attacked, but there is no one to help. Twelve-year-old Michelle also drew her brother, whom she described as sad because his parents (not hers!) have been taken away. She has nightmares in which her parents go to court, they are deported and she is unable to support her brother, so he dies. Keydy, a 12-year-old seventh grader whose father has six years and whose mother has 12 years of formal education, drew her person as a soccer goalie (initially without a face) and identifies with him as "the goalie on the losing Spanish team who doesn't give up." She worries about parental deportation, and she has nightmares about being killed at an airport and her parents crying. The windows of her apartment house are empty of curtains or mullions, suggesting a perceived lack of privacy.

As we have seen, the usual HTP norms do not always hold with this population. A chimney need not be present for the household to be experienced as warm and loving, and drawing a person of the opposite sex need not mean gender dysphoria. In fact, one may seriously misconstrue the meaning of any child's drawing if no narrative is sought or no cultural context is not considered.

A feminist approach to our evaluations and the data generated has proved helpful in maintaining an awareness of these cultural and gender-driven differences. It has informed our observation that the chronic stress of living "under the radar" combined with familial separation and high levels of aspiration tend to increase with chronological age, cognitive capacity and the immediate pressure of a parental attempt to change immigration status.

Furthermore, it appears that the situational stress experienced by these girls so overwhelms most gender differences that it creates a situation in which gender does not seem to be a central organizing axis of power for this cohort. In fact, as we reexamined the data, it became clear that neither the quantitative nor qualitative data from the interviews showed any difference between the boys and the girls, with only the one exception included above. Girls may be more sensitive to these situational stressors, but this was not reflected in our sample.

Indeed, family fragmentation and the loss of safety that is created by the threat of separation/removal were almost inevitably associated with emotional distress, even when the current separation was only from the grandparents. These observations support Maslow's (1943) theory of a hierarchy of needs which asserts the primacy of the need for safety over all other psychological needs, as well as its further development in attachment theory as elaborated by Ainsworth and Bowlby (1965). We further assert that girls with greater intelligence and accelerated cognitive development display a greater appreciation of their situation and therefore suffer greater stress earlier.

Conclusion

A full psychological hardship evaluation documents the usefulness of these immigrants to this society and the *extreme and unusual hardship* to which their daughters would be subject if a parent were deported. In the context of appropriate legal representation, it can be an effective tool for obtaining official residential status for the parents and, by extension, for family and child safety, security and membership in our society. In this way it aids redress of the underlying power imbalance and contributes to ameliorating the harmful effects of their daughter's second-class citizenship. The extent of this harm is so great as to overwhelm the gender differences one would expect to observe. Current immigration policy regarding change of status proceedings paradoxically tends to favor those cases in which the citizen child is physically, psychologically and/or cognitively compromised. However, our evaluations document the significant pathogenic effects of familial separation and/or forced deportation upon the potential *extreme and unusual hardship* of girls who are physically healthy, cognitively age-advanced and scholastically successful. Although apparently functioning well, the threat of potential deportation and/or parental loss and familial fragmentation is emotionally destructive for high functioning girls at an earlier age than for their peers, and this threat is greater than generally acknowledged, because they have a greater awareness of the implications of these outcomes. The current greater likelihood that these girls will suffer deportation is truly ironic because they would be among those best able to make a positive contribution to our society. Their narratives illuminate a challenge to US immigration policy.

We call upon researchers to further explore the experiences of such citizen daughters (and sons) and to more fully document their distress. This will enhance our ability to implement prevention, early intervention and supportive services for these first-generation US citizens. Such work will enhance their developmental potential, empower their citizenship and support their capacity to contribute to society.

References

Ainsworth, M. & Bowlby, J. (1965). *Child care and the growth of love*. London: Penguin.

American Psychological Association (2007). Guidelines for psychological practice with girls and women. *American Psychologist, 62*, pp. 949–979. doi:10.1037/0003–066X.62.9.949.

American Psychological Association (2013). Specialty guidelines for forensic psychology. *American Psychologist, 68*, pp. 7–19. doi:10.1037/a0029889.

American Psychological Association Presidential Task Force on Immigration (2012). *Crossroads: The psychology of immigration in the new century*. Washington, DC: Author.

Brady-Amoon, P. (2011). Humanism, feminism, and multiculturalism: Essential elements of social justice in counseling, education, and advocacy. *Journal of Humanistic Counseling, 50*, pp. 135–148.

Buck, J.N. (1946, 1969). *House tree person manual*. Torrance, CA: Western Psychological Services.

Caplan, P.J. (1995). *They say you're crazy: How the world's most powerful psychiatrists decide who's normal*. Reading, MA: Addison-Wesley.

Capps, R., Castaneda, R.M., Chaudry, A. & Santos, R. (2007). *Paying the price: The impact of immigration raids on America's children*. Retrieved from http://www.urban.org/UploadedPDF/411566_immigration_raids.pdf.

Chaudry, A., Capps, R., Pedroza, J.M., Casteneda, R.M., Santos, R. & Scott, M.M. (February 2010). *Facing our future: Children in the aftermath of immigration enforcement.* Retrieved from http://www.urban.org/uploadedpdf/412020_FacingOurFuture_final.pdf.

Chesler, P. (1972/2005). *Women and madness.* New York, NY: Palgrave Macmillan.

Crawford, M. (2013). Meaning making and feminist practice. *Psychology of Women Quarterly, 37,* pp. 256–260.

Espin, O.M. (1987). Psychological impact of migration on Latinas: Implications for psychotherapeutic practice. *Psychology of Women Quarterly, 11,* pp. 489–503.

Espin, O.M. (1997). *Latina realities: Essays on healing, migration, and sexuality.* Boulder, CO: Westview.

Espin, O.M. (1999). *Women crossing boundaries [electronic resource]: A psychology of immigration and transformations of sexuality.* New York: Routledge.

Esquivel, G.B., Oades-Sese, G.V. & Olitzky, S.L. (2008). Multicultural issues in projective assessment. In L.A. Suzuki & J.G. Ponterotto (Eds.), *Handbook of multicultural assessment: Clinical, psychological, and educational applications* (3rd ed.) (pp. 346–372). San Francisco: Wiley.

Devore, J.E. & Fryrear, J.L. (1976). An analysis of juvenile delinquents; hole drawing responses on the tree figure of the House-Tree-Person technique. *Journal of Clinical Psychology, 32,* pp. 731–736.

Feminist Therapy Institute (1999). *Feminist therapy code of ethics.* Retrieved from http://chrysaliscounseling.org/Feminist_Therapy.html.

Garcia, A. (2013). The facts on immigration today. *Center for American Progress.* Retrieved from http://www.americanprogress.org/issues/immigration/report/2013/04/03/59040/the-facts-on-immigration-today-3/.

Harari, R. & Harar, H. (2006). Children's environment and health in Latin America. *Annals of The New York Academy of Sciences, 1076,* pp. 660–677. doi:10.1196/annals.1371.082.

Hondagneu-Sotelo, P. (Ed.) (2003). *Gender and U.S. immigration: Contemporary trends.* Berkeley, CA: University of California Press.

Human Rights (2012). *Ecuador Country Review,* pp. 56–59.

Ingoldsby, B., Schvaneveldt, P., Supple, A. & Bush, K. (2004). The relationship between parenting behaviors and adolescent achievement and self-efficacy in Chile and Ecuador. *Marriage & Family Review, 35,* pp. 139–159. doi:10.1300/J002v35n03_08.

Kalyan-Masih, V. (1976). Graphic representation: From Intellectual realism to visual realism in a Draw-a-House-Tree Task. *Child Development, 47,* pp. 1026–1031.

Kirschner, S.R. (2013). Diagnosis and its discontents: Critical perspectives on psychiatric nosology and the DSM. *Feminism & Psychology, 23,* pp. 10–28. doi:10.1177/0959353512467963.

Koppitz, E.M. (1968). *Psychological evaluation of children's human figure drawings.* New York: Grune & Stratton.

Kraul, C. (June 26, 2011). Drug trafficking on the rise through Ecuador. *Los Angeles Times.* Retrieved from http://articles.latimes.com/2011/jun/26/world/la-fg-ecuador-drugs-20110626.

Larrea, C. & Kawachi, I. (2005). Does economic inequality affect child malnutrition? The case of Ecuador. *Social Science & Medicine, 60,* pp. 165–178. doi:10.1016/j.socscimed.2004.04.024.

Leeds, A., Dirlam, D. & Brannigan, G. (1983). The development of spatial representation in children from five to thirteen years of age. *Genetic Psychology Monographs, 108*, pp. 137–165.

Maslow, A.H. (1943). A theory of human motivation. *Psychological Review, 50*, pp. 370–96.

Morrow, S.L., Hawxhurst, D.M., Montes de Vegas, A.Y., Abouselman, T.M. & Castaneda, C.L. (2006). Toward a radical feminist multicultural therapy. Renewing a commitment to activism. In R.L. Toporek, L.H. Gerstein, N.A. Fouad, G. Roysircar & T. Israel (Eds.), *Handbook for social justice in counseling psychology. Leadership, vision, and action* (pp. 231–247). Thousand Oaks, CA: Sage.

Ogdon, D.P. (1981). *Psychodiagnostics and personality assessment: A handbook* (2nd ed.). Los Angeles: Western Psychological Services.

Orcés, D.M. (2009). Democratic values and public opinion toward immigrants: The case of Ecuador. *Latin American Politics & Society, 51*, pp. 131–155. doi:10.1111/j.1548–2456.2009.00066.

Padilla, A.M. & Borsato, G.N. (2008). Issues in culturally appropriate psychoeducational assessment. In L.A. Suzuki & J.G. Ponterotto (Eds.), *Handbook of multicultural assessment: Clinical, psychological, and educational applications* (3rd ed.). San Francisco: Jossey-Bass.

Pessar, P. (2003). Engendering migration studies. The case of new immigrants to the United States. In P. Hondagneu-Sotelo (Ed.), *Gender and U.S. immigration: Contemporary trends* (pp. 20–42). Berkeley, CA: University of California.

Political Conditions (2012). *Ecuador Country Review*, pp. 10–29.

Rae-Espinoza, H. (2010). Consent and discipline in Ecuador: How to avoid raising an antisocial child. *Ethos, 38*, pp. 369–387. doi:10.1111/j.1548–1352.2010.01156.x.

Rivera, L. (2010). Acculturation: Theories, measurement, and research. In J.G. Ponterotto, J.M. Casas, L.A. Suzuki & C.M. Alexander (Eds.), *Handbook of multicultural counseling* (3rd ed.) (pp. 331–341). Thousand Oaks, CA: Sage.

Rogers, C.R. (1980). *A way of being*. New York: Houghton-Mifflin.

Safran, S. (1997). *The H-T-P View Point*, 2. Nyack, NY: Author

Short, E.L., Suzuki, L., Prendes-Lintel, M., Furr, G.P., Madabbushi, S. & Mapel, G. (2010). Counseling immigrants and refugees. In J.G. Ponterotto, J.M. Casas, L.A. Suzuki & C.M. Alexander, *Handbook of multicultural counseling* (3rd ed.). (pp. 201–212). Thousand Oaks, CA: Sage.

Stutman, G. & Brady-Amoon, P. (2011). Supporting dependant relatives of undocumented immigrants through psychological hardship evaluations. *Journal of Forensic Psychology Practice, 11*, pp. 369–390. doi:10.1080/15228932.2011.583909v.

Thorne, B., Orellana, M.F., Lam, W.S.E. & Chee, A. (2003). Raising children, and growing up, across national borders. Comparative perspectives on age, gender, and migration. In P. Hondagneu-Sotelo (Ed.), *Gender and U.S. immigration: Contemporary trends* (pp. 241–262). Berkeley, CA: University of California.

U.S. Census (2003). *The foreign-born population*: 2000. Retrieved from http://www.census.gov/prod/2003pubs/c2kbr-34.pdf.

U.S.Census (2010). Table 1.1. *Population by Sex, Age, Nativity, and U.S. Citizenship Status: 2010*. Retrieved from http://www.census.gov/population/foreign/files/cps2010/T1.2010.pdf.

Wilkinson, S. & Kitzinger, C. (2013). *Representing our own experience: Issues in "insider" research. Psychology of Women Quarterly, 37*(2), pp. 251–255.

Western Psychological Services (2013). The House-Tree-Person. Retrieved from http://portal.wpspublish.com/portal/page?_pageid=53,70613&_dad=portal&_schema=PORTAL.

Yost, M.R. & Chmielewski, J.F. (2012). Blurring the line between researcher and researched in interview studies: A feminist practice? *Psychology of Women Quarterly, 37*, pp. 242–250. doi:10.1177/0361684312464698.

11

Intergenerational Narratives of Migration and Acculturation: *Ni de aquí ni de allá*

Sandra Mattar and María Teresa Pestana

"We want our children to become world citizens." These words were uttered by my Lebanese immigrant parents when I (Sandra) was an early teenager struggling with issues of identity in Venezuela, my native country. Little did I know that those words would have a significant impact on my life's journey as an immigrant in the United States a journey that was very much tied to my family's particular circumstances, but also to the cultural, historical and sociopolitical context of the countries that represent my family's journey. An analysis of the complex intersection between psychosocial adjustment and acculturation embedded in all these factors is necessary to understand the way my daughter (María Teresa) is growing up and acculturating in the US. Otherwise, the migration story is only half-told.

According to Piper (2005), there is a sharp increase in global mobility and resettlement around the world. Women represent the majority of these international migrants. At the same time, transnationalism has contributed to complex patterns of acculturation and resettlement. All these factors underscore the need to understand migration narratives across generations and the impact of multiple contextual factors on acculturation processes. Our current knowledge on the psychology of migration and acculturation is still basically one-dimensional, and it is described in the psychology literature as "one size fits all" to refer to the uniform description of immigrants' acculturation processes regardless of the immigrant classification, gender, country of origin and racial and ethnic background, among others (Rudmin, 2003). Acculturation is described in very linear and simplistic terms, ignoring the multiple pathways that cultural identity can take (Jensen, Arnett & McKenzie,

2011) as well as the gendered, racialized and politicized experiences of immigrants (Espín, 1999; Bhatia & Ram, 2009; Boyd, 2012; Green, 2012). There is also a need for more clarity on the role of social class, sexual orientation and language use, among others, in determining the outcomes of these experiences (Bhatia & Ram, 2009; Agbényiga & Huang, 2012). More importantly, it is now recognized that gender is "a central organizing principle in migration flows and in the organization of migrant's lives" (Lutz, 2010) and that there is a need to study gender in its multiple contexts in order to avoid essentialism (Stewart & McDermott, 2004).

This chapter will address some aspects of the gendered immigration and acculturation gap in the literature by including the immigration narratives of two generations of women impacted by voluntary transnational serial migration. The first author, first of all, will explore her immigration experience as the white daughter of Lebanese immigrants to Venezuela and, second, will reflect on her Lebanese mother experiences as an immigrant herself. She will then look at her shifting experience as an American citizen Latina and as the mother of three "American-born" children, using a critical self-reflection model (McWhirter, 1994) and an intersectionality perspective (Crenshaw, 1995). Finally, the chapter will include the main author's American daughter's narrative around her experience of embodying four different ethnic groups and her searching for a "core" identity. The authors' main goal is to understand the multiplicity of contextual factors that shape some women's immigration experiences and to elucidate some aspects of the intergenerational effects of migration on women. In this sense, Chamberlain & Leydesdorff (2004) state that "memories and personal accounts remain a unique, and often the only, way to access migrant experience, sensitivities and identities" (p. 228). The authors will discuss their multiple negotiations of self-identity, gender roles, professional roles, empowerment, cultural pride and motherhood. A special emphasis will be placed on the phenomenon so close to the intergenerational immigrant experience: *"ni de aquí ni de allá"* experience (neither from here nor from there). The authors will use the existing literature on intersectionality, acculturation, internalized racism and colonial trauma to frame their lived experiences. By rethinking and discussing the multiple pathways of gendered immigration, the authors hope to contribute toward feminist migration analysis and theory. We also hope to further our current (and limited) understanding of acculturation theories by describing some of the contextual and demographic forces that have influenced our acculturation process.

Sandra's story

I am the daughter of Catholic (Maronites) Lebanese immigrants to Venezuela and am now a US citizen. My parents immigrated to Venezuela at different ages due to "pull factors" or a life of opportunities in a country where they heard Lebanese countrymen where doing very well economically. My parents arrived in that country during one of the most significant developments in Venezuelan history, the dictatorship of General Marco Pérez Jiménez (1952–1958), who modernized that country at a remarkable pace. No other time in Venezuelan history has there been such a remarkable progress or such a welcoming stance toward immigrants. In fact, many immigrants from Europe arrived in Venezuelan shores escaping the poverty of World War II. These immigrants were well educated, and they found jobs awaiting them. Venezuelans opened their arms to these European newcomers. But this was not the first time Europeans were welcomed in the country. Due to unfortunate remnants of internalized colonialism, there were some previous attempts by the government during President Guzmán Blanco's administration (mid-1800s) to "improve the race" by offering land and jobs to German immigrants to entice them to move to Venezuela. This is how "La Colonia Tovar" was born, a German agricultural enclave made up of peasants from the Baden region of Germany and located in the middle of the mountainous region outside of Caracas, the capital. The descendants of these German immigrants are still found there, although some have returned to Germany.

There were other immigration waves to Venezuela thereafter, such as the immigration of upper-middle-class white Cubans who were escaping Castro's regime. In the early eighties, other immigrant groups arrived from Colombia due to the oil bonanza that hit Venezuela as a result of the government nationalizing the oil companies. A rush of cheap laborers invaded the country. Local reception of these immigrants was very tepid, unlike the one demonstrated after the European and original Cuban immigration.

Lebanese immigrants in Venezuela.

According to Schwartz, Unger, Zamboanga & Szapocznik (2010) the acculturation process is not a linear one, nor is it only defined by the choices made by migrants. Therefore, it is important to understand the interactional context of the immigrants in order to understand their acculturation pathways. This context includes the characteristics of the countries of origin and the receiving countries as well as the characteristics of the migrants, including their race, socio-economic level, gender, sexual orientation and ability to speak the language of the country in which they would resettle.

When many Lebanese immigrants first arrived in Venezuela in the late 1800s, they carried a Turkish passport (a product of the Ottoman Empire's domination of much of the Middle East). Carrying a Turkish passport defined my parent's nationality and the way they were perceived and nickname by locals. These immigrants were known among Venezuelans as "los turcos" (the Turkish ones). This nickname was very humiliating for my parents. The Turks dominated Lebanon for several years and established a brutal and ruthless domination. In their efforts to spread their feudal regime, they persecuted and tortured many Lebanese Christians, and they desecrated local churches and massacred ten thousand people in 1860 (Kisirwani, 1980). By the time my parents moved to Venezuela, they brought with them a legacy of traumatic memories from the Turks' invasion, a French colonial influence, two languages (French and Arabic) – which made learning Spanish very easy – one religion (Catholic Maronite) and an ambition to "explore and trade" with the locals inherited from their ancestors, the Phoenicians, great explorers and seamen, and the creators of the alphabet. Lebanese immigrants overall brought with them a serious work ethic and a remarkable ambition. They also brought the color of their skin. Many of them also brought a French education, which was a product of French domination and colonization in Lebanon. Unbeknownst to them, both race and education proved to be very advantageous to their social mobility among the Venezuelans.

At the time of my parents' arrival in Venezuela in the 1950s, the country was ruled by a white majority that strongly favored European background and perpetuated the social dynamics inherited from "la época de los mantuanos", a rigid social-class hierarchy, which was the product of the Spanish domination of the country. It is not a coincidence, then, that my family encountered a similar dynamic of social classes ruled by European colonial values as well as a strong belief in their Catholic religion. It was a match made in heaven, and one that was very successful. My parents chose European names for their children: Edwin, Emily, Sandra and Astrid, thereby perpetuating European dominant values. While the Venezuelan dominant class baptized their children using Spanish colonial names, my parents used more Nordic, British and French names to distinguish us from the rest of the locals in Venezuela. My mother went along with my dad's wishes to give us those European names. Her family members were "sheiks" or members of the

Maronite nobility in Lebanon, which is the product of a feudal system still reigning in that country. Many of them were leaders of villages – in many cases, government officials. They were all well educated and spoke French fluently. European influence was pervasive, so the names they gave us were consistent with that influence.

In my parents' minds, these names were conveyors of success, sophistication and class. It was also their way to distinguish us from the rest of the locals. This act was partly a way to convey a superior cultural heritage while at the same time positioning themselves among the elite in Venezuela.

The gender roles that my parents brought with them were also consistent with the Venezuelan local context: The oldest son ruled over his siblings, and his needs had priority over his female siblings. Women were seen but not heard. It was privately understood that they were active participants in family decisions, but their role was never publicly acknowledged as such. My parents' education played a role in loosening these social mores somehow, albeit in a paradoxical way. My father was a believer in empowering his daughters to achieve whatever they wished to do. However, when it came to the family business, it was clear that my brother would become the inheritor of the business, as well as get, by far, the biggest share of the family's inheritance. We received a strong (yet subtle) message that, as women, our power was "limited".

The similarities between the two cultures around collectivism and gender roles (Schwartz et al., 2010), coupled with my parents' work ethic, French influence and skin color, as well as the local welcoming context, played a significant role in my parents' psychosocial adaptation and acculturation. These intervening factors might also explain the widespread success that many Lebanese immigrants experienced in Venezuela.

Growing up as "the other"

Growing up in Venezuela, I always grappled with the idea of who I was and with finding my place among my Venezuelan peers. Like my parents before me, I was also nicknamed "la turca". However, unlike my parents, I had very different reasons for not liking this nickname. This nickname made me different and categorized me as "the other" or as "non-Venezuelan". Upon reaching my late teen years, I chose the path of identification with my parents' Lebanese background. Partially following the "separation" strategy of acculturation proposed by Berry (1980), I joined a Lebanese youth group, and I even became the editor of a magazine that romanticized everything Lebanese, and provided a forum for Lebanese Venezuelan youth to voice and share different perspectives related to their lives and identities.

My parents were very proud of their cultural heritage, and they made sure their children inherited this culture. I therefore grew up in a Lebanese bubble where Lebanese values, costumes and traditions prevailed. Our weekends were taken with outings with members of my family and of the Lebanese diaspora. Lebanese poets and singers filled my life and a sense of nostalgia for the country "that was". Arabic was the language mainly spoken at home. As we grew up, French was "de rigueur" as part of our education. Along came the French songs and the French novels. All of these were part of my Lebanese Venezuelan education. Interestingly enough, I went to an all-girl's Catholic school that was run by nuns from Spain. In the process, I was quickly internalizing the colonial values of the two countries. I was part of the dominant group without being aware of it and was quickly becoming the "world citizen" that my parents so wished to see in their children.

Dating in Venezuela was a complicated matter. Following similar patterns among immigrants around the world wanting to recreate the costumes of the culture that they left (Sluzki, 1979), my siblings and I were each socialized to marry a Lebanese "candidate". However, like many immigrant children growing up in a different culture from that of their parents, our interests extended beyond the Lebanese reality. We were interested in dating Venezuelans against our parents' advice. Two of us ended up marrying Venezuelans of other backgrounds, while the remaining two married a man and a woman, respectively, of Lebanese descent.

After marrying a Venezuelan man of Portuguese descent, I headed to the US to "follow him". My husband had a clear plan of what he wanted to do: attend graduate school in Boston, Massachusetts. I had no plans of my own except for supporting him. In this sense, while my immigration to the US was by choice, I did not consider my own needs when "deciding" to move to the US. However, I decided not to adopt my husband's last name. Unbeknownst to me, this decision would have significant ramifications in the development of my sense of identity in the US.

Immigration to the US: From a white "turca" to a brown Latina woman

Moving to the US forced me to examine who I was and affirm my identity as a woman. This is why I decided to keep my family name and not adopt my husband's last name. While I knew this would be a big blow for both my mother and my mother-in-law, I kept my family name and, with it, a big part of my identity. I broke free from expected societal traditions to adopt my husband's name. I am almost convinced that had I remained in Venezuela, I would have been subjected to traditional patriarchal arrangements and therefore would have changed my surname, no questions asked.

According to Espín (2010), "becoming a member of a new society stretches the boundaries of what is possible.... [T]he crossing of borders through migration provides the space and 'permission' to cross emotional boundaries and transform their gender roles" (Espín, 2010). Deciding to keep my name was an empowering act that would pave the way for other life decisions in the US, such as attending graduate school and becoming the first Latina associate dean in the history of the college where I currently work. Unknowingly, with that act, I sealed the pattern of interaction with my husband and a renegotiation of the gender role in our marriage and in our life in the US. This experience is consistent with a statement from Suarez-Orozco & Qin (2006): "migration can challenge expectations about gender-related roles requiring renegotiations" (p. 170). Likewise, Harris, Firestone & Vega (2005) indicate that "the process of acculturation may be accompanied by changes in family dynamics, including expectations about appropriate role behavior for men and women" (p. 464).

Paradoxically, while I became more empowered with regard to my gender roles, I also had to experience the parallel pressures of living up to the expectations of a Venezuelan mother: devoted to her husband, her children and her household. My family back in Venezuela made sure that I was fulfilling these roles (via phone, Internet or other means). My husband struggled with his beliefs in gender equality and the need to have a traditional wife at his side. Therefore, becoming a professional in the US and a working mother forced me to balance the roles of a professional, an American working mother and a Venezuelan mother. This balancing act resulted in some psychological and emotional costs for me and for my marriage.

While I have to deal with many adaptations upon moving to the US, nothing prepared me for the blow to my identity and my sense of privilege that I was about to experience as a Latina immigrant in Boston, as well as for the experience of acculturation stress. In fact, as Espín (1997; 1999) and Escobar, Nervi & Gara (2000) indicate, immigrant women experience a host of stressors that significantly affect their health and well-being, including uprootedness; nostalgia; cognitive overload of learning a new language and a new system; loneliness; low self-esteem; questioning their level of competence, appearance and self-worth; and cultural mourning (Ainslie, 1998). These factors become more salient when immigrants have minority status in the host country.

Becoming "the other" in the US by way of national origin and accent forced me to examine who I was, as well as the privileges I had benefited from as an upper-middle-class white woman in Venezuela. This turned out to be the most significant and spiritual experience in my life so far – as well as the most stressful and painful. Like many immigrants facing new cultural expectations, I had to negotiate dominant, non-dominant and ethnic groups' expectations.

"Ni de aquí ni de allá": Negotiating a new identity

As the daughter of Lebanese immigrants in Venezuela, I struggled with the idea of who I was (both in Venezuela and in the US). My parents' rapid social mobility and positioning among the upper-middle-class society exposed us to a system that accepted us because we embodied "the right" race, class and education. However, we could not trace back our ancestors to Spain, and this did not make us "real Venezuelans". I have carried this sense of not being a "real" citizen with me to the United States. My insecurities related to my sense of belonging (or lack thereof) is always "triggered" when people in the US ask me, "where is home?" after I've lived here for 25 years. I suppose I also carry the legacy of colonialism inherited by my parents and also learned in Venezuela as well as the sensibilities that are the product of being perceived as "the other".

Intersectionality, race and gender

Immigrant women's experiences can vastly differ from those of their male counterparts (Espín, 1999, 2006). In fact, Mahalingam, Balan & Haritatos (2008) stress the need for using an intersectional and gendered analysis of immigration psychology in order to differentiate the experiences of immigrants. Furthermore, some authors stress the importance of making immigrant service delivery gender-specific (Agbényiga & Huang, 2012). Intersectionality theory stresses that gender stands at the crossroads of multiple variables, such as race, class, sexual orientation and ethnic background, among others, as well as the multiple interactions between these variables, thereby resulting in a qualitatively different identity than when examined alone: the greater is more than the sum of the parts. The success of negotiating these multiple identities will depend on multiple contextual factors, such as the level of oppression in the host country and available support systems (Yakushko & Chronister, 2005).

While I was growing up with multiple identities, I never questioned or examined my power as an upper-middle-class white woman. My urgency for identity was not around skin color but around national origin. In this sense, white race did not define my identity, because this privilege was unexamined. Ethnicity had more prevalence. My ancestors did not trace back to Venezuela; therefore, I was not "the real thing", as someone told my Venezuelan nephew recently. Like many immigrants, I faced the pressures to adopt the Venezuelan cultural heritage practices but not the identifications (Schwartz et al., 2010).

It was not until I came to the US that I lost my status as a white woman and all the privileges that came with it. These circumstances lead me to embrace my Venezuelan heritage with fervor. With my Spanish accent and my Venezuelan passport, I was now a "real Venezuelan" to Americans around me. We even gave Venezuelan names to our children in the US. My Lebanese identity moved to the background. Political developments in the Middle East and American perception of immigrants from that region of the world made it difficult for me to share my Lebanese background. I found myself being treated as having the "right-race-but-not-really", the "wrong" ethnicity (Latina), the "wrong" accent (Spanish) and a confusing (in essentialist terms) combination of all: a (privileged) white immigrant woman with a Latino accent, a Venezuelan-Lebanese heritage, a doctoral degree in psychology and a polyglot. I represented an amalgam of stereotypes and identities that would influence my life, my children's lives, my self-esteem and my psychosocial adjustment in the US in various ways. This experience will influence my rejection of essentialist and pure descriptions of identity vis-à-vis immigrants that place them "in opposition to an authentic 'native' identity" as well as gender essentialism that homogenizes the experience of Latinas (Mahalingam & Leu, 2005).

The complexity of the intersection of race, gender and context is amplified when I interact with some Latina/os in the US. They see me as a Venezuelan, white and upper-middle-class woman, with all the privileges that come with that. I am placed in a Latin-American context within the US. This recreation of the old social order is one more way that challenges natives' essentialist representations of my gender, race, class and ethnicity, and it underscores the need to use an intersectional and contextual approach when referring to immigrant women's acculturation process (Mahalingam & Leu, 2005).

María Teresa's story

"What are you?" Every child of mixed racial and cultural background who has ever taken a standardized test in America has at least paused

before deciding which bubble to fill in. As a child struggling to wrap my mind around the concept of race, its application and its consequences, my uncle would urge me to fill in "Other" and in the blank space write, "World Citizen". This worked to bolster my sense of identity during the early years of grade school by embracing my transnational roots. But as time passed I lost pride in my global heritage and instead became disillusioned with the fact that there would never be a bubble to fit me: American, Venezuelan, Lebanese, Portuguese.

The other

Throughout my life I have been constantly asked, "where are you from?" As a child, this question would confuse me. After replying with, "I was born in Boston, Massachusetts," a follow-up question always ensued: "No, where are you *really* from?" The way this second question was posed, even though at the time I did not know its meaning/the answer, meant somehow that I was not from here. Apart from being born in America, I had American friends and went to American schools and lived in America. What other explanation could I give to "prove" I was American? Growing up as a brown child of immigrants in a predominantly racially and culturally white neighborhood, I began to understand early on that I was different from my peers.

The broader societal context in which immigrants and their children find themselves is of great importance because it signals the ease or difficulty of their adaptation. As one of the only Spanish-speaking families in our monolingual neighborhood, it was automatically assumed that we were from Mexico. Given that Mexicans are the leading ethnic group of Latina/os in California and the US (United States Census Bureau, 2010), it is not so unusual that I be grouped in a predominant ethnic group. Although annoyed by the constant misidentification (both of my parents are from Venezuela), I quickly saw how the identification of "Mexican" was less of ethnic and national significance but more of racial stereotypes and stigma (Ortiz & Telles, 2012). The running "joke" was that my dad mowed lawns and my parents didn't know how to speak English and were poorly educated or were illegal migrants. The fact that both of my parents held doctorate degrees, spoke several languages and were well traveled was disregarded. Not only does the stigma surrounding Mexicans continue to permeate my peers' perspective of my family; it also perpetuates the disparaging attitudes toward this vibrant group, dismissing the important contributions they make to American society. The dissonance between who my parents were and how others saw them deeply influenced my path toward self-identification.

Balancing act

As a second-generation immigrant, I have experienced the growing pains of becoming an American. Because of their new environment, children of immigrants must strike a balance between the demands of their tradition-oriented parents and also the demands of American culture. The cultural socialization processes of immigrants no longer have the same value to their children because the host country often does not share the same cultural values, beliefs or practices of the immigrants' culture. Therefore, meaningful connections to the "old" world are often lost or diluted, and a disconnect forms between the reality of the "new" world their children grow up in and this "old" world (Zhou, 1997).

Not only was I my parents' first child, but through me, my parents experienced their many "firsts", as well. Their acculturation experiences were learned second-hand mainly through my first-hand experiences in school, which is arguably the most influencing means of American socialization. Because my parents were learning through me, I was convinced that I understood the intricacies of American culture and society and that my parents did not. So in my eyes, I was different because my parents were different. More apparent to me was the fact that my parents were different from others; they had accents, dressed funny and were stricter than my classmates' parents. I constantly asked myself why I couldn't be a kid with "normal", "American" parents who talked, dressed and acted "right". It was because I saw my parents through the eyes of my peers – as foreign and strange – that I yearned to be accepted by my peers.

This desire to fit in meant distorting and ignoring my and my family's cultural distinctiveness. Ultimately, I understood that fitting in would never occur. After visiting my parents' home country several times since childhood, I developed strong ties with my family abroad and a meaningful connection to the land and its people – a sort of symbolic citizenship. Yet certain aspects of my identity did not translate so easily. Speaking Spanish evolved from a means of communicating with my family to an indicator of our "otherness". When accompanying my parents on public outings in the US I would demand that they not speak to me in Spanish for fear that someone might hear us. When my parents would pick me up from a friend's house, I secretly hoped that they would not stay for too long or talk too much – or at all – to my friend's parents. I felt stuck in between the dissonant worlds of my family and of America while trying to understand who I was. As Suarez-Orozco & Qin (2006) indicate, "when there is too much cultural dissonance, when cultural guides are inadequate, and when the social mirror reflects negative images, adolescents may find it difficult to develop a flexible and adaptive sense of self" (p. 177). Given this internal dilemma, it is no surprise that I have struggled with feelings of self-acceptance and not belonging for many years.

At times I have become depressed with the disconnect I feel from others. Even among some of my friends, I feel like there is a level of knowledge and identity experience that they will never understand, simply because they are not in my unique position of embodying so many cultures, as well as being a minority in general and feeling like an outsider.

Internalized racism and stereotype threat

The pressures surrounding social stereotypes about my family deeply influenced my academic behavior. While signing up for the SAT in high school, I could not bring myself to select my high school as the test location, where all of my friends and classmates would be taking it. Without fully being aware of my decision, I selected a school I had never heard of because its name appealed to me and I knew the city's diversity would be reflected in its schools. On test day I was surprised but comforted to see the overwhelming number of Latina/o, black and Asian students as well as the dearth of white students. It was then when I realized that I truly was a minority in my environment at school and in my community. Rarely had I seen so many students my age who were minorities. Although I knew not a single person, I felt comfortable and confident. I fit in. But somehow I felt more prepared than these students. In retrospect, not only was I trying to avoid the "stereotype threat" (Steele and Aronson, 1995); I simultaneously carried with me those stereotypes about minorities, or what some scholars call internalized racism (Padilla, 2001). I could not take a test of such high stakes among people that I thought were smarter and better than me, so I went somewhere where I knew I would feel more prepared. It boosted my self-esteem.

Finding my own identity

Having different ethnic identities has allowed me to gain access into the various social groups that are attached to them. However, I feel I'm on the fringe of each group: "ni de aquí, ni de allá." I have learned that when asked, "where are you from?" the answer that people want to hear is not, "I was born in Boston and moved to California when I was four years old": it is, "my parents are from Venezuela, and my grandparents are from Lebanon and Portugal." Like filling in the race bubble on a standardized test, there is a tendency for others to want to define who

I am. Yet my identity is not so static. Today, the way I define myself is contingent upon the social location I find myself in. I still struggle with feelings of inferiority and positive self-image and with the idea of who I am. I may not feel confident with my identity in every social context, but I welcome the multitude of perspectives it has allowed me to take on: as brown, as white, as a woman and as an outsider.

Parenting in English and raising María Teresa (Sandra's voice)

Raising children in a foreign country without the support of one's family is probably one of the most stressful and isolating experiences immigrant women can have. This is also probably the case for many women hailing from collectivistic cultures. The burden of motherhood in a new country is heightened by others stressors encountered by the immigrant woman, such as discrimination, pressures to adopt local child-rearing cultural practices, balancing home and work, acculturation stressors, reorganization of gendered roles in the marriage, brain overload and sheer physical exhaustion, among others.

The expression of my innate need to transmit my culture to my children and my own way of parenting felt stumped through a process whereby I was scrutinized by locals around issues of questioning (or not) of parental authority by the children, overt physical expressions of love and affection, allowing dependent behaviors vs. encouragement of self-reliance, discipline style, and level of monitoring children's lives. Ochoka & Janzen (2008) call these beliefs and values that immigrant parents want to transmit to their children "parenting orientation", and they call the parenting changes as parents adapt in the host country "parenting modifications". These modifications depend on several factors, such as the length of resettlement in the host country, the strength of the immigrants' cultural and religious values, speed of adaptation in the new country and economic resources of the family, among others (Ochoka & Janzen, 2008). I would also add to this list the examination of the bicultural values I brought with me and the internal divisions I experienced as a result of having to choose my parenting orientation and having to choose from those cultural values I wanted to transmit (Venezuelan, Lebanese and American). The emotional and psychological cost of having to constantly navigate and question these assumptions, or the erosion of the ethos and social and cultural capital (Waters, 1997; Yosso, 2005), have been difficult for me. Furthermore, life decisions around where to live or where to send the children to school also had a significant impact on our children's self-esteem and on my sense of accomplishment as a parent. We learned later on in our parenting experience of the hazards of raising children in public school districts where they represented the one percent minority in a vastly white school (both students and teachers). Our major challenges were how to position our children for success in an environment that stereotyped (see McDermott & Varenne, 1995) and invalidated them while it afforded them resources and a "valued" mainstream education. Just like with many immigrants and underprivileged groups in mainstream society (and unlike my experience in Venezuela as a white child), our children became victims of the politics of exclusion in their school and community, such as through a lack of representation of their culture in school curricula, gross stereotypes regardless of their parents' educational and socio-economic levels, assumptions about their academic abilities and questioning their American identity. The result of this politics of exclusion and deficit theories (Yosso, 2006) is usually internalized racism, lack of confidence, stereotype threat (Steele & Aronson, 1995) and low self-esteem. Unfortunately, educators frequently ignore the consequences of exclusion because of "inequity, disinvestment and marginalization" in the system (Arzubiaga, Nogueron & Sullivan, 2009, p. 252).

The research literature on immigrant children (Portes & Rivas, 2011) suggests that variables such as socioeconomic status, access to good schools, parents' high educational level and English proficiency are protective factors for immigrant children. However, many of these studies fall short in describing the complexities of factors that determine a positive identity and acculturative experience for these children, including the particular sociocultural environment in which families settle and the local political tensions affecting new immigrants (Arzubiaga et al., 2009). In my family's case, regardless of our social and cultural capital and our educational level (both parents are college professors), we had limited influence vis-à-vis the discriminatory practices and politics of exclusion in our children's schools. At the same time, we have transmitted to them the cultural values and traditions of four different cultures, along with two languages. This cultural capital will stay with them forever.

Summary

This chapter explored the intersection of gender, country of origin, languages, class, race, colonialism, previous trauma histories and acculturation processes, among others, in determining the lived experiences of three generations of female immigrants impacted by serial intergenerational migration. Using their own immigration narratives, the authors argued that the experience of immigration can be understood through only the cultural, historical and sociopolitical contexts of the countries that represent the authors' family journey through three different continents. Specifically, the narratives illustrate the need to particularize each immigrant's story, and the dangers resulting from theoretical generalizations on the immigrant's experience. Likewise, there are themes that emerge from the narrative around the need for self-definition and identity, the need for place and the psychological sequelae resulting from being an outsider. This chapter has highlighted the multiple paths that acculturation and identity can take and the dangers of the use of simplistic models to understand the immigration experience. They recommend the use of an intersectionality framework to account for a more complex and dynamic understanding of women's migration and acculturation experiences.

References

Agbényiga, D.L. & Huang, L. (2012). Gendered Immigration: Implications and Impact on Social Work Education. *Advances in Social Work, 13*(2), pp. 291–305.

Ainslie, R. (1998). Cultural mourning, immigration, and engagement: Vignettes from the Mexican Experience. In M.M. Suárez-Orozco (Ed.), *Crossings: Mexican Immigration in Interdisciplinary Perspectives*. Cambridge, MA: David Rockefeller Center for Latin American Studies. Harvard University Press, pp. 283–300.

Arzubiaga, A.E., Noguerón, S.C. & Sullivan, A.L. (2009). The education of children in im/migrant families. *Review of Research in Education, 33*, 246–271.

Berry, J.W. (1980). Acculturation as varieties of adaptation. *Acculturation: Theory, models and some new findings*, pp. 9–25.

Bhatia, S. & Ram, A. (2009). Theorizing identity in transnational and diaspora cultures: A critical approach to acculturation. *International Journal of Intercultural Relations, 33*(2), pp. 140–149.

Boyd, M. (August 2012). Destination Canada (?): Gender, race and changing policies for recruiting migrant domestic workers. In *The Second ISA Forum of Sociology (August 1–4, 2012)*. Isaconf.

Chamberlain, M. & Leydesdorff, S. (2004). Transnational families: Memories and narratives. *Global Networks, 4*(3/2004), pp. 227–241. ISSN 1470–2266.

Crenshaw, K. (1995). *'The intersection of race and gender': Critical race theory: The key writings that formed the movement.* New York: New Press.

Escobar, J.I., Nervi, C.H. & Gara, M.A. (2000). Immigration and mental health: Mexican Americans in the United States. *Harvard Review of Psychiatry, 8*(2), pp. 64–72.

Espín, O.M. (1997). *Latina realities: Essays on healing, migration, and sexuality.* Boulder, CO: Westview Press.

Espín, O.M. (1999). *Women crossing boundaries: A psychology of immigration and transformation of sexuality.* New York: Routledge.

Espín, O.M. (2006). Gender, sexuality, language, and migration. In R. Mahalingam (Ed.), *Cultural psychology of immigrants* (pp. 241–258). Mahwah, NJ: Erlbaum.

Espín, O.M. (2010). *Is domestic violence a cultural tradition?...and other questions about gender and migration.* Presentation at the United States Embassy (Vienna, June, 2010).

Green, N.L. (2012). Changing paradigms in migration studies: From men to women to gender. *Gender & History, 24*(3), pp. 782–798.

Harris, R.J., Firestone, J.M. & Vega, W.A. (2005). The interaction of country of origin, acculturation, and gender role ideology on wife abuse. *Social Science Quarterly, 86*(2), pp. 463–483. doi:10.1111/j.0038-4941.2005.00313.x.

Jensen, L.A., Arnett, J.J. & McKenzie, J. (2011). Globalization and cultural identity. In *Handbook of identity theory and research* (pp. 285–301). New York: Springer.

Kisirwani, Maroun (October 1980). "Foreign interference and religious animosity in Lebanon". *Journal of Contemporary History, 15*(4), pp. 685–700. doi:10.1177/002200948001500405. JSTOR 260504.

Lutz, H. (2010). Gender in the migratory process. *Journal of Ethnic and Migration Studies, 36*(10), pp. 1647–1663.

Mahalingam, R. & Leu, J. (2005). Culture, essentialism, immigration and representations of gender. *Theory & Psychology, 15*(6), pp. 839–860. doi:10.1177/0959354305059335.

Mahalingam, R, Balan, S. & Haritatos, J. (2008). Engendering immigrant psychology: An intersectionality perspective. *Sex Roles, 59*, pp. 326–336. doi:10.1007/s11199-008-9495-2.

McDermott, R. & Varenne, H. (1995). Culture "as" disability. *Anthropology & Education Quarterly, 26*, pp. 324–348.

McWhirter, E.H. (1994). *Counseling for Empowerment.* Alexandria, VA: American Counseling Association.

Ochoka, J. & Janzen, R. (2008). Immigrant parenting: A new framework of understanding. *Journal of Immigrant & Refugee Studies, 6*(1), pp. 85–111.

Ortiz, V. & Telles, E. (2012). Racial identity and racial treatment of Mexican Americans. *Race and Social Problems*, 4(1), pp. 41–56.

Padilla, L.M. (2001). But you're not a dirty Mexican: Internalized oppression, Latinos, and law. *Texas Hispanic Journal of Law & Policy, 7*, pp. 59–113.

Piper, N. (2005). Gender and migration. *Policy Analysis and Research Programme. Geneva: Global Commission on International Migration.*

Portes, A. & Rivas, A. (2011). The adaptation of migrant children. *The future of children, 21*(1), pp. 219–246.

Rudmin, F.W. (2003). Critical history of the acculturation psychology of assimilation, separation, integration, and marginalization. *Review of general psychology, 7*(1), p. 3.

Schwartz, S.J., Unger, J.B., Zamboanga, B.L. & Szapocznic, J. (2010). Rethinking the concept of acculturation: Implications for theory and research. *American Psychologist, 65*(4), pp. 237–251.

Sluzki, C.E. (1979). Migration and family conflict. *Family process, 18*(4), pp. 379–390.

Steele, C. (1997). A threat in the air: How stereotypes shape intellectual identity and performance. *American Psychologist, 52*(6), pp. 613–629.

Steele, C.M. & Aronson, J. (1995). Stereotype threat and the intellectual test performance of African-Americans. *Journal of Personality and Social Psychology, 69*, pp. 797–811.

Stewart, A.J. & McDermott, C. (2004). Gender in psychology. *Annual Review of Psychology, 55*, pp. 519–544.

Suarez-Orozco, C. & Qin, D.B. (2006). Gendered perspectives in psychology: Immigrant origin youth. *International Migration Review, 40*(1), pp. 165–198. doi:10.1111/j.1747-7379.2006.00007.x.

U.S. Department of Commerce, United States Census Bureau (2010). *Nation's foreign-born population nears 37 million* (CB010–159). Retrieved from http://www.census.gov/newsroom/releases/archives/foreignborn_population/cb10–159.html.

Waters, M. (1997). Immigrant families at risk: Factors that undermine chances for success. In A. Booth, A.C. Crouter & N. Landale (Eds.), *Immigration and the Family: Research and Policy on U.S. Immigrants* (pp. 79–87). Mahwah, NJ: Lawrence Erlbaum.

Yakushko, O. & Chronister, K.M. (2005). Immigrant women and counseling: The invisible others. *Journal of Counseling and Development, 83*, pp. 292–298.

Yosso, T. (2005). Whose culture has capital? A critical race theory discussion of community cultural wealth. *Race, Ethnicity and Education, 8*(1), pp. 69–91.

Zhou, M. (1997). Growing up American: The challenge confronting immigrant children and children of immigrants. *Annual Review of Sociology, 23*, pp. 63–95.

12
Negotiating Identity for Young Adult Women from Immigrant Families: Expectations, Opportunities and Challenges

Karen Kisiel Dion

Coming to terms with the personal meaning of one's ethnicity can be challenging during late adolescence and early adulthood for daughters born in the society to which their parents immigrated. Given these challenges that I'll discuss at the start of this paper, I am interested in how Canadian-born university undergraduate women whose parents immigrated to Canada understand and think about their ethnocultural heritage. Given the demands and pressures of gendered socialization within immigrant families across diverse ethnocultural communities, some young women may de-emphasize and/or redefine the role of their ethnocultural heritage in their self-definition. In the second part of this paper, based on comments from interviews with undergraduate women from immigrant families, I propose that this approach does not represent denial of one's roots but may, in some circumstances, reflect an adaptive approach to constructing one's identity.

Growing up in an immigrant family is an experience shared by many individuals in Canada, especially in large metropolitan areas. Canada is a major immigrant-receiving society, particularly in, though not limited to, large urban centers. As of the 2006 census, nearly one-fifth of those who were enumerated were born outside Canada (Statistics Canada, 2006). Toronto continues to be Canada's "major immigrant gateway" (Statistics Canada, 2006) with nearly 46 percent of the Toronto census metropolitan area born outside of Canada in 2006. This census area includes not only the city of Toronto but also many of the surrounding suburban areas. There is strong social structural support for multiculturalism, which has been part of Canadian national policy since the early 1970s (see Reitz, 2009 for a discussion of the history of multiculturalism in Canada and elsewhere). One important feature of this policy is recognition of heritage cultures and support for the maintenance of traditions including heritage language, festivals and other events. A second feature is ensuring equal opportunity in various domains of public life.

Gendered socialization in immigrant families

Families who immigrate seek new opportunities in the receiving society but also hope to protectively keep what is precious to them from their society of origin. What is precious often is about ways of being in close relationships, particularly within the family – customs and traditions around important life events, parent-child interaction and responsibilities of family members. Gender frames and guides expectations and decisions about both preserving what is important that immigrants derived from their society of origin and adapting within the receiving society. Daughters' experiences differ from sons' experiences in many domains. Compared to sons, daughters often are expected to exemplify traditional gender-related attitudes and behaviors considered desirable and appropriate in the parents' society of origin. There is considerable evidence that in immigrant families there is greater parental monitoring of daughters' compared to sons' behavior (see Dion & Dion, 2001). Several factors underlie greater parental and family regulation of daughters' behavior.

First, beliefs about optimal identity development and the role of parents and other family members in a young adult's development are culturally constructed. In various classic models of identity development (Erikson, 1968; Marcia, 1980), the family's role is one of "developmental facilitation" (see Dion, 2006). The developmental task for the young adult is to explore various options and develop their own sense of personal identity through experiences within and outside the family, as illustrated more recently by the concept of "emerging adulthood" (Arnett, 2000). The concept of "identity foreclosure" (Marcia, 1980) as contrasted with "identity achievement" refers to forming one's sense of identity based on others' influence rather than one's individual reflections and choice. However, the belief that parents should support and facilitate their young adult children's independence, foster their children's ability to make their own decisions and lessen their children's dependence on parents and other family members reflects a series of cultural assumptions about the nature of family. From the perspective of cultural analyses, scholars have argued that fostering psychological autonomy from one's family of origin is not the aim of socialization in many societies (see LeVine, 1990; Shweder & Bourne, 1984). From this perspective, parents in many immigrant families may expect to be actively involved in decision-making across diverse domains of their young adult children's life. For the reasons discussed above, daughters are likely to be the focus of more parental attention and involvement in their lives than are sons.

Greater monitoring of daughters' compared to sons' behavior is attributable to not only a desire to preserve important values and behaviors from the family's society of origin. Feminist scholars have persuasively argued that the symbolic importance of a daughter's behavior has implications well beyond its personal meaning for her and her parents. Le Espiritu (2001) suggested that among various immigrant groups, a sense of "moral superiority" over those with more power in the receiving society could be claimed through their daughters' adherence to traditional patterns of behavior. Thus greater parental restrictiveness reflected both a desire to protect core values and a desire to assert parents' "political and sociocultural claims on their new country" (Le Espiritu, 2001, p. 415). From this perspective, women from immigrant families are critical to the community's "cultural representation". Similarly, Espín (2006) noted that the behavior of one's daughter(s) is regarded as an indicator of the moral standards of the family in many cultural contexts; thus how a daughter behaves has implications for the family and its reputation. In addition, Espín (2006) also noted that the experience of immigration itself requires considerable adaptation and can often mean losing control over many aspects of one's life. She noted that one way of regaining some sense of control is by continuing to expect that family members, especially daughters, will adhere to gender-related norms of behavior.

These family expectations can be associated with conflict in several ways. Parents' expectations formed at the time they immigrated may no longer be representative of current ways of family functioning in their society of origin (e.g. Ghosh, 1984). If so, there may be disagreement among family members, such that the nature of cultural tradition becomes an "embattled category" (Das Gupta, 1997, p. 580). Behaviors which are functional in school or work contexts (e.g. assertiveness) may be associated with conflict when they occur with parents or other relatives if these behaviors challenge expectations of respect and authority (e.g. Pettys & Balgopal, 1998).

At the same time, family-related values, practices and behaviors from their parents' society of origin can also be a source of pride for young adult children. Among the young adult children from immigrant families, both feelings of tension and satisfaction may coexist in the context of family relationships (Dion, 2006). Thus for daughters from immigrant families, the expectations of parents and other relatives also contribute positively to their sense of identity and role in the family. The above discussion raises the question how these gendered expectations about family-related behavior and behavior with peers contribute to young women's identity development.

Daughters' reactions to gendered socialization in immigrant families

To explore this understanding, we, myself and (the late) Ken Dion undertook an in-depth interview study of young adult women from immigrant families. In much previous research on gender and cultural identity, the focus has been on issues related to a specific ethnocultural group. While we aim not to minimize the importance of this approach, our own and others' research on the social psychology of immigration has found that immigration raises gender-related issues pertaining to one's identity both within and across diverse groups. The research I'll be discussing in the remainder of this paper involved detailed interviews with 147 women (Canadian university undergraduate students) from several ethnocultural communities: both Asian (Chinese, Indian and Pakistani) and European (Italian and Greek). Most were in their first year of university, with an average age of 19 years. They were all single, and most lived at home with their parents and other family members. These interviews were conducted in the late 1990s. All of the participants were born in Canada. Their parents were immigrants to Canada. Interviews always took place first, followed by completion of several structured measures.

Material from the interviews will be discussed here. The interview questions were from an interview protocol developed by Phinney and her colleagues (see Phinney & Devich-Navarro, 1997) to examine how adolescents from different ethnocultural groups come to terms with the meaning of their cultural backgrounds, especially their feelings about different aspects of their identity. Each participant was interviewed individually. All interviews were taped with the participant's consent and transcribed. Considered together, the questions examined participants' feelings and thoughts about their ethnocultural background, as well as their feelings and thoughts about being Canadian. There were also a series of questions designed to address different aspects of being bicultural. As originally designed by Phinney and her colleagues, the interview protocol focused on culture-related questions, including valued aspects, problems and sources of conflict. Interestingly, although gender was not explicitly mentioned in the initial overview of the questions or in the specific questions asked, participants often mentioned gender-related issues. A major source of disagreement between Canadian-born daughters from immigrant families, on one hand, and their parents and other family members, on another, centered on parental monitoring and regulation of daughters' behavior, especially regarding socializing with peers as the following example illustrates:

> most people here will go out in the evening, and ... they'll come home late but back there, you know you're supposed to be home early. ... You know how ... you're kind of bad if you come home late and stuff? ... Plus I'm a girl so it's even harder. ... Certain things like that, I disagree with."

Consistent with Le Espiritu's points discussed previously, there is a sense of contrasting expectations of appropriate behavior for "most people here" compared to "back there", namely her parents' home before coming to Canada and an awareness of the negative judgments of the latter when daughters stay out late.

Casual interaction between women and men similarly elicited intergenerational disagreement, once again with reference to various social contexts:

> I sometimes disagree with what my parents might think about ... male-female relations. ... They think that ... males and females shouldn't be too close. ... But ... I don't see what the problem is. ... You can just be ... casual friends. ... And obviously in this society, you have to work ... with people of the opposite sex.

Another related theme of disagreement between daughters and their parents centered on dating and eventual choice of a spouse. A daughter's desire to follow the cultural script of dating as a way of getting to know about potential partners often conflicts with a very different view of dating on the part of parents and other family members as follows:

> When you have a boyfriend they think: oh, no no no no. You shouldn't have a boyfriend until you are ready to get married ... how

> am I supposed to know the person? Or how am I supposed to develop certain feelings?

As mentioned earlier, there are very different cultural models of what defines good parenting. Given the perspective previously discussed on the symbolic importance of daughters' behavior for the family, it might be expected that these models are particularly strong and clearly defined for parents' socializing their daughters as these comments illustrate:

> At my home I'm supposed to just accept…whatever decision my parents do so that has been a big problem. Usually it's not really solved because I…have to…act out whatever my dad's decision was…but I'm hoping to wear him down as I try to get my own way as well and try to expose him to the other attitudes of Canadian culture.
>
> I think it should be up to the individual to choose whom he or she wants to marry and who she wants to spend the rest of their life with but, apparently, not any of the elders or my parents agree.

Finally, conflict over expected gender roles was evident in some interviews: for example,

> I disagree…many times.…The importance that they place on…marriage…even if a woman does go to university…and gets an education, once she gets married…and has children, she's supposed to…just hang her diploma up on the wall and forget everything and take care of her family.

Many of the points of disagreement or conflict revolve around degree of control over personal life, including nature of family obligations and participation, choice of friends, regulation of time spent outside the family, restrictions on dating and cross-sex interaction. Given these issues, it can be asked what impact these challenges have on daughters' self-definition, in terms of their thoughts and feeling about their own identity. The women interviewed in many cases lived in very different cultural worlds at home as contrasted with interactions with peers in school and work contexts. LaFromboise, Coleman and Gerton (1993) proposed that it was possible for an individual to develop "bicultural competence" through a set of specific skills which would enable her or him to take part in different cultural contexts. These skills included culturally relevant knowledge and communication abilities, developing appropriate roles and social ties in each cultural context, resulting in a "sense of being grounded" and of particular relevance here, "bicultural efficacy", the feeling that one can live biculturally given relevant skills that allow successful alternation between different cultural contexts. They identified two aspects of identity development – namely "self-sufficiency" (a secure and strong sense of oneself) and a sense of one's cultural identity – and suggested that being bicultural might be a positive experience with distinct psychological benefits for the individual. Birman (1994) distinguished between the behavioral aspect of biculturalism and its impact on psychological factors, such as identity, and stressed the importance of considering both aspects. She suggested that these two features of biculturalism might function independently. Various accounts of bicultural identity suggest diverse forms (see LaFromboise et al., 1993; Birman, 1994; Phinney & Devich-Navarro, 1997; Benet-Martinez & Haritatos, 2005; Schwartz, Unger, Zamboanga & Szapocznik, 2010). Kankesan (2010) found across three different studies of Canadian university undergraduates on an ethnically diverse campus that given several options for self-description, most of these young adults saw themselves as bicultural rather than focusing either only on Canadian identity or on their identity rooted in their ethnocultural heritage/ancestry.

The distinction between behavioral competence and a sense of psychological identity discussed previously is important here. Taking part both in family life and in life as a university student indicated bicultural behavioral competence, with individual differences in some domains (e.g. heritage language ability). In most of their homes, the women described their parents as maintaining specific culture-related traditions and practices. When these women were asked how they felt about maintaining these traditions, the typical response was positive, ranging from commenting it was fine or not minding to strong enthusiasm. Only a few replies were clearly negative or ambivalent. Reading through the narratives, what emerged from many of them was a sense of appreciation of the positive contributions their ethnocultural background had on their family and personal life, combined with, in some cases, trying to develop their own sense of what it meant to be a good daughter given important points of disagreement between themselves and family members.

Given this context, an important issue is how these undergraduate women from immigrant families described their own sense of identity, in particular whether they felt more strongly identified with their ethnocultural heritage, with being Canadian or with both of these possibilities. The most frequent response was to describe oneself as bicultural: Canadian and the specific cultural background of their family. The next most frequently occurring response was to identify more strongly as Canadian. In this case, it often occurred that the participant would then point out that choosing this option didn't imply a distancing from her heritage but instead a result of being born and raised in Canada – hence greater familiarity and experience in this context. For those who felt bicultural, they were asked to choose among three options, using the interview protocol developed by Phinney & Devich-Navarro (1997) mentioned earlier: namely, a sense of alternating between two identities depending on the situation, a sense of combining both identities and thus being a mixture of the two or not explicitly thinking of oneself with reference to culture but "to think of yourself in other ways". As expected, some participants endorsed feeling a sense of a blended identity, while others described themselves as alternating between two cultures in their daily lives.

Interestingly and unexpectedly, however, a substantial number of these university undergraduate women who felt both Canadian and also felt a connection to their cultural heritage chose the third way "to be part of two cultures", that is, not explicitly thinking about themselves in cultural terms in their ongoing lives.

Here are a few examples:

> I usually do ... what my heart tells me to do. ... I mostly look at myself as a person ... I don't really look at it much as a cultural value or something I have to do because it's a part of my culture. I do it because it's for myself.
>
> I don't sit there and think: ... how would my parents' traditional views ... affect my decision? ... I do things for myself. ... when I'm interested in something, I'll just do it, regardless of any cultural implications.
>
> I think of myself in other ways. Always with somewhere rooted in me ... my culture or cultures. But when I make decisions ... it's more based on me and my morals, values ... but on me and not on my culture.
>
> I think culture is something rather subconscious. Like it's part of you even though you don't realize it ... I'll just say I'm ... a person ... everybody has a different personality no matter what culture you come from ... I'm a nice person and funny. I have a good sense of humor. I can be intelligent.
>
> I don't every day wake up and think about my culture. It's just sort of part of me ... I think of myself as being ... a woman. ... I think more

> about ... my personality ... am I a kind person? Am I ... a mean person? ... I tend to focus on what I have control over ... and the person that I'm becoming ... what I'm doing right now.

The greater-than-expected prevalence of this third choice was intriguing. What is the meaning(s) of this "choosing to think of oneself in other ways?" For most of the women who chose this option, this choice did not seem to represent denying an important aspect of oneself, but rather a reframing and reinterpretation of the contribution of one's cultural heritage. Psychologically, this reframing might help to mitigate felt pressures to follow a series of gendered cultural prescriptives. By not self-categorizing themselves with a group label, a broader range of options for identity and self-evaluation are available. As noted earlier, gendered socialization expectations and practices were a source of disagreement between parents and their daughters. At the same time, many of the participants reported a positive involvement in practices and traditions related to their heritage culture. Perhaps the above approach permits a delicate psychological balancing between these two daily realities. As previously mentioned, ethnocultural diversity is a very salient feature of their daily life, as is the experience of immigration, both in terms of the city where they live and as students on a university campus where many students are from families whose parents immigrated to Canada.

The comments from the interviews seem to reflect a positive sense of "taking for granted" the benefits of this diversity, both in terms of the perceived advantages of being born and raised in Canada and in terms of growing up with, for the most part, active family observance of cultural practices/traditions from other societies. The comments cited earlier reflect an awareness of differences in gendered expectations between their families' and their own views on family obligations; behavior outside the family, especially with peers; attitudes toward dating; and marriage. Negotiating bicultural identity for many of these women seemed to be more like setting out on a complex but adventurous trip rather than embarking on a perilous journey. Whichever path they were following on this trip, the narratives, for the most part, suggested a clear sense of agency and of active involvement and positive engagement in dealing with the differences between themselves and their family. In a few cases where the narratives indicated a high level of gender-related conflict, the lack of any sense of agency seemed to be a major issue.

This emphasis on personal agency expressed by these young women is consistent with some evidence from research with young adults from immigrant families. Dion and Dion (2004) found, using the "Aspects of Identity Measure" developed by Cheek and his colleagues (Cheek & Tropp, 1994; Cheek, Tropp, Chen & Underwood, 1994), that Canadian-born university students from immigrant families regarded personally distinctive features as more important for their sense of identity compared to those born elsewhere who came to Canada with their families. Compared to men from immigrant families, women from immigrant families regarded this aspect of identity as more important. These features included psychological states such as "my thoughts and feelings", "my self-knowledge", "my ideas about what kind of person I really am" and "my feeling of being a unique person, being distinct from others". By contrast, another finding from this study was that young adults who were themselves first-generation immigrants born outside Canada reported a more interdependent view of self-emphasizing a sense of connectedness between oneself and others (using the measure developed by Singelis, 1994) compared to their Canadian-born peers from immigrant families.

The most intriguing finding that emerged from the present study was the choice of the third option for being bicultural, namely thinking of oneself in other ways among some of those interviewed. This finding suggests that for some second-generation women from immigrant families, individuality and self-determination can be grounded in a feeling of cultural connection without explicitly thinking of oneself in cultural-group-related terms. Their comments acknowledged the importance of their cultural background as a factor in shaping their individuality rather than viewing it as providing a set of rules that guided their ongoing decision making. It is important to stress that the choice of a third option did *not* represent distancing themselves from their heritage culture or from their identity as Canadian. This choice was not a sign of alienation, judging by comments made by most of these participants to questions about their interest in cultural background and degree of participation, but rather a preference for thinking of themselves either without any label or in terms of their personality or interests. These women's narratives suggest a potentially important direction for understanding women's bicultural identity – namely the factors contributing to and the implications for identity negotiation when cultural factors are an implicit as contrasted with an explicit part of self-definition. As discussed at the start of this paper, gender frames important family dynamics within many immigrant families. Given gender-related expectations and pressures to embody specific patterns of behavior and cultural-related roles for second-generation women, thinking of oneself "in other ways" offers a way of addressing these pressures. The distinction between biculturalism as behavioral and feelings of personal identity as discussed by Birman (1994) is worth re-emphasizing here. The women in this research were for the most part at the beginning of their university education and living at home with their respective families. Many important life events were still ahead of them. It can be argued that this approach may function as a form of empowerment at the psychological level while continuing to maintain and develop close relationships at the behavioral level within and beyond one's family.

References

Arnett, J.J. (2000). Emerging adulthood: A theory of development from the late teens through the twenties. *American Psychologist, 55*, pp. 469–480.

Benet-Martinez, V. & Haritatos, J. (2005). Bicultural identity integration (BII): Components and psychosocial antecedents. *Journal of Personality, 73*, pp. 1015–1050.

Birman, D. (1994). Acculturation and human diversity in a multicultural society. In E.J. Trickett, R.J. Watts & D. Birman (Eds.), *Human diversity: Perspectives on people in context* (pp. 261–284). San Francisco: Jossey-Bass.

Cheek, J.M. & Tropp, L.R. (1994). *The aspects of identity questionnaire: History and bibliography*. Unpublished manuscript. Available from the first author, Department of Psychology, Wellesley College, Wellesley, MA.

Cheek, J.M., Tropp, L.R., Chen, L.C. & Underwood, M.K. (1994). *Identity orientations: Personal, social, and collective aspects of identity*. Paper presented at the meeting of the American Psychological Association, Los Angeles, CA.

Das Gupta, M. (1997). "What is Indian about you?" A gendered, transnational approach to ethnicity. *Gender and Society, 11*, pp. 572–596.

Dion, K.K. & Dion, K.L. (2001). Gender and cultural adaptation in immigrant families. *Journal of Social Issues, 57*, pp. 511–521.

Dion, K.K. & Dion, K.L. (2004). Gender, immigrant generation, and ethnocultural identity. *Sex Roles, 50*, pp. 347–355.

Dion, K.K. (2006). On the development of identity: Perspectives from immigrant families. In R. Mahalingam (Ed.), *Cultural psychology of immigrants* (pp. 299–314). Mahwah,NJ: Erlbaum.

Erikson, E.H. (1968). *Identity: youth and crisis*. New York: Norton.

Espín, O.M. (2006). Gender, sexuality, language, and migration. In R. Mahalingam (Ed.), *Cultural psychology of immigrants* (pp. 241–258). Mahwah, NJ: Erlbaum.

Ghosh, R. (1984). South Asian women in Canada: Adaptation. In R. Kanungo (Ed.), *South Asians in the Canadian mosaic* (pp. 145–155). Montreal, Quebec: Kala Bharati.

Kankesan, T. (2010). *Understanding bicultural identity and its impact on the association between discrimination and well-being*. Unpublished doctoral dissertation, University of Toronto, Toronto, Canada.

LaFromboise, T., Coleman, H.L.K. & Gerton, J. (1993). Psychological impact of biculturalism: Evidence and theory. *Psychological Bulletin, 114*, pp. 395–412.

Le Espiritu, Y. (2001). "We don't sleep around like white girls do": Family, culture and gender in Filipina American lives. *Signs, 26,* pp. 415–440.
LeVine, R.A. (1990). Enculturation: A biosocial perspective on the development of self. In D. Cichetti & M. Beeghly (Eds.), *The self in transition: Infancy to childhood* (pp. 99–117). Chicago: University of Chicago Press.
Marcia, J.E. (1980). Identity in adolescence. In J. Adelson (Ed.), *Handbook of adolescent psychology* (pp. 159–187). New York: Wiley.
Pettys, G.L. & Balgopal, P.R. (1998). Multigenerational conflicts and new immigrants: An Indo-American experience. *Families in Society: The Journal of Contemporary Human Services, 79,* pp. 410–423.
Phinney, J.S. & Devich-Navarro, M. (1997). Variations in bicultural identification among African American and Mexican American adolescents. *Journal of Research on Adolescence, 7,* pp. 3–32.
Reitz, J.G. (2009). Assessing multiculturalism as a behavioural theory. In J.G. Reitz, R. Breton, K.K. Dion & K.L. Dion (Eds.), *Multiculturalism and social cohesion: potentials and challenges of diversity* (pp. 1–47). Dordrecht, NL: Springer.
Schwartz, S.J., Unger, J.B., Zamboanga, B.L. & Szapocznik, J. (2010). Rethinking the concept of acculturation: Implications for theory and research. *American Psychologist, 65,* pp. 237–251.
Schweder, R.A. & Bourne, E.J. (1984). Does the concept of the person vary cross-culturally? In R.A. Schweder & R.A. LeVine (Eds.), *Essays on mind, self and emotion* (pp. 158–199). Cambridge: Cambridge University Press.
Singelis, T.M. (1994). The measurement of independent and interdependent self-construals. *Personality and Social Psychology Bulletin,* 20, pp. 580–591.
Statistics Canada (2009). *Immigration in Canada: A Portrait of the foreign-born population, 2006 census.* Catalogue no. 97–557. Retrieved from http://www12.statcan.gc.ca/ census-recensement/2006/as-sa/97–557/pl-eng.cfm.

13

Legacies of Migration: Italian American Women, Food and Identity

Andrea L. Dottolo and Carol Dottolo

Carol: Do you think there is a more authentic or Italian version [of this recipe for "Italian meatballs"]?

Donna: Hmmm, well I'm sure today there could be. Only because of everybody watching the fat content. And trying to eat healthier, you know?

Carol: So you think it would be, they might modify the ingredients?

Donna: Yeah, I'm not sure Italians would do it, but maybe German. Or, you know, Irish might use like, turkey instead of... [hamburger].

Carol: Ugh! It gives me chills!

Carol, Paula & Donna: [laughter].

Donna: I know! Me too!

This quote is an excerpt from an interview that is part of a study on Italian American women, identity and food, which is the focus of this chapter. The interviewer is Carol, whom you will hear more about later, and she is in a room with the interviewee, Donna, and Donna's friend Paula. (All names of interviewees are pseudonyms). I begin this chapter with the words of one of the women in order to set the stage, to introduce and contextualize this study. The central themes of this study are authenticity and Italian imaginary.

This chapter explores the following research questions: How do Italian American women understand and imagine their "Italianness", sometimes several generations removed from migration? How do women use food as part of intergenerational maintenance of ethnic identity, and connection to an (often imagined) immigrant legacy? How do Italian American women use food to understand themselves as "real" Italians and as different from women who are not of Italian ancestry?

At this point, the focus is on the questions about the recipe that appeared toward the middle of the interview:

> Do you ever make substitutions for the ingredients in this recipe? Is there a more "authentic" or "Italian" version of this recipe?
>
> What feelings or emotions are associated with eating this food/dish? What feelings or emotions are associated with making or preparing this food/dish?

When Carol asks if there is a more authentic version of the recipe for the Italian meatballs that is being discussed, Donna considers the question and responds, "hmmm, well I'm sure today there could be." By stating that "there could be", Donna communicates that she doesn't really know, indicating an imagined, constructed standard of authenticity. She also points out that "today" there could be another version of the recipe, distinguishing the present from the past, quite possibly an imagined immigrant past.

Donna postulates that this alternative (Americanized) recipe is likely lacking "because of everybody watching the fat content". She is clear that "real", authentic Italians would not modify the ingredients, but non-Italians – that is, inauthentic others, explicitly referring to Germans or the Irish and implicitly referring to German-*Americans* and Irish-*Americans* – might. Since "traditional" Italian meatball recipes include hamburger, veal and/or pork, but never turkey, Donna's mere mention of the poultry ingredient immediately incites horror in Carol ("Ugh! It gives me chills!"), followed by the laughter of recognition in all who are present. Donna joins in the group cohesion with the exclamation, "I know! Me too!" by agreeing that turkey in meatballs is disgusting. This excerpt illustrates the sharing of insider status, of an "authentic" Italian Americanness, and how food can function to mark outsider status – in this case, a non-Italian "other".

While there have been significant contributions about Italian American women in the fields of history, sociology and literature (e.g. Guglielmo, 2010; Bona, 1999; Giunta, 2002; di Leonardo, 1987; Gabaccia & Iacovetta, 2002), very little research has been conducted from a psychological perspective. Given the paucity of scholarship in psychology as a whole about Italian American women, there is an even greater absence of literature in psychology interpreted through a feminist lens focusing on this population. This chapter focuses on granddaughters of immigrants and on how they construct and maintain their ethnic identity – in this case, being Italian. Since language cannot serve as symbol for these women, because speaking Italian was often forbidden in their homes, or was spoken between only adults in covert communications, they often must cling to other symbols of Italianness in order to preserve their sense of gendered ethnic identities. I argue that one such symbol is food, wherein participants manipulate recipes and food to navigate and negotiate being both Italian and American, femininity, social class and honoring both a lost and imagined cultural experience.

[Why does the identity of Italian American need to be preserved? After all, these women are American.... This is paradoxical because they are Americanized, and contrary to the assumed desire to second- and third-generation immigration of losing the old country, or the reality that they have already lost it... maybe just point out that it is interesting to note that while there is the assumption they would want to lose, they don't, and they do it through food].

DeSalvo and Giunta (2002) explore the relationship between Italian American women and food, claiming that food is connected to "gender, sexuality, ethnicity, race, class, power, the environment, health, immigration, politics and culture", while

> contextualized within broader aspects of Italian American life and culture, such as birth, motherhood, family relationships, death, generational changes, gardens, rituals, and sexual identity, but also within aspects of Italian history, such as immigrant history; the condition of the peasantry in Italy; one's relationship to Italy, the country left behind; and forms of homecoming. (p. 3)

By focusing on the words of Italian American women, I hope to further illuminate the claims that DeSalvo and Giunta (2002) made regarding the multiplicity and complexity of food in the life and identity of Italian American women.

Social identities

This study is situated in social identity theory, where scholars often credit much of the groundwork in the field to American psychologist Henri Tajfel (1981), who described social identity as "that part of an individual's self-concept which derives from his [*sic*] knowledge of his [*sic*] membership of a social group (or groups) together with the value and emotional significance attached to that membership" (p. 255). It is group membership that informs, constructs and maintains social identity. Jenkins (1996) offers a similar characterization: "Social identity is our understanding of who we are and of who other people are, and, reciprocally, other people's understanding of themselves and of others (which includes us)" (p. 5). Group membership and group relations serve as interpretive signifiers, directing social navigation. This is especially relevant when considering how Italian American women understand their own Italianness in relation to others who are both similar to and different from them.

However, ethnic or national identity cannot be examined in a vacuum. Feminist theories of intersectionality inform us that our multiple social identities intersect and are mutually constitutive, such that I cannot examine gender without also considering race and class, for example (Dottolo & Stewart, 2008; Crenshaw, 1995; Combahee River Collective, 1977).

Marecek (1995) provides an illustrative example of intersectionality, highlighting its significance to scholars of social identity:

> To say that I am a White, middle-class, North American woman is not to say that I am part woman, part White, and part late 20th century North American. If I were fitted out with skin of a different color, or transported into a different moment of history, I would no longer be the woman I am now. (Marecek 1995, p. 163)

Maracek not only demonstrates the importance of including multiple identities and institutional structures in our analyses but also highlights how our insights might be lacking, that our appreciation for narratives of identity will inherently exclude or ignore potentially significant dimensions in capturing a fuller, more accurate account.

Imaginary

Sociologist Richard Jenkins illustrates an important psychological component of social identities as located primarily in the minds of individuals, claiming that "Although... identities are imagined[,] they are not imaginary" (1996, p. 175). Characterizing identities as "imagined" depicts them as social constructions, though indelibly "real". While all identities are "imagined" in this sense, for immigrants and their descendants, preserving the home culture and traditions is necessarily invented, constructed, remembered and misremembered about a place and time that no longer exists. As these anchors of the culture are told and retold, practiced and diluted over time and thus become part of selective memory in a new and imagined "home".

In other words, for Italian American granddaughters of immigrants, their social identities are created and recreated through historical interpretations. Their identities are historical responses to immigration. Scott (2001) asserts,

> Categories of identity we take for granted as rooted in our physical bodies (gender and race) or our culture (ethnic, religious) heritages are, in fact, retrospectively linked to roots; they don't follow predictably or naturally from them. (Scott 2001, p. 285)

That is, our collective and individual histories continuously mold and shape who we think we are. In an exploration of the relationship between histories and social identities, Scott (2001) invokes a term naively created by an undergraduate student as an analytical intervention. Scott remembers that as a graduate teaching assistant she graded undergraduate student essays and came across a mistake. In an attempt to capture a European professor's heavily accented pronunciation of a French term describing the period at the end of the nineteenth century, or "fin de siècle", the student wrote it as "fantasy echo". Scott (2001) employs this concocted English phrase as a strategy for investigating the historicity of social identities. In deconstructing the term "fantasy echo", Scott posits,

> Depending on whether the words are both taken as nouns or as an adjective and a noun, the term signifies the repetition of something imagined or an imagined repetition. In either case the repetition is not exact since an echo is an imperfect return of sound. Fantasy, as noun or adjective, refers to plays of the mind that are creative and not always rational. For thinking the problem of retrospective identification it may not matter which is the noun and which is the adjective. Retrospective identifications, after all, are imagined repetitions *and* repetitions of imagined resemblances. The echo is a fantasy, the fantasy an echo; the two are inextricably intertwined. (p. 287)

The fantasy echo, then, becomes a kind of metaphor that is particularly relevant for the psychology of immigration. If the sounds represent the memories of culture, tradition, place and space in the home country, then that sound over time bounces off of hard surfaces – trauma, institutional structures, loss, longing and nostalgia. The sound crashes against the new culture, region, language and relationships, and it is propelled, warped or disrupted by class, race, gender, sexuality, age and disability. Finally, the sound reverberates back, often to new listeners, interpreting these new sounds in a new way.

Louise DeSalvo (2004) eloquently describes the imaginary in her journey to return to Italy to rediscover her immigrant grandparents.

> I will never know my grandparents' histories. No matter how much I try to remember my grandparents' stories; how much I listen to my father's stories; how much I read; how much I study of my grandparents' photos, their papers, the few artifacts that remain of their lives. No matter whether or not I visit the places they lived.
>
> All I can do is conjecture, imagine, invent their lives. My story of their story, a distortion, a misrepresentation of what they lived. But my story of their story, now a respect I must accord them – though I cannot possibly get it right, though I cannot possibly understand who they were – so their lives do not pass into oblivion. (p. 143)

DeSalvo explains her many attempts to piece together a history, a connection, an experience that she knows is mostly lost. But the vestiges that she has left, woven into a story, are what ground her to her family and Italianness. For Italian American women, having lost so much of the immigrant culture, food becomes the vessel through which to hold on to the culture, history and identity.

Authenticity

As the fantasy echo sounds over generations of women who identify as Italian American, establishing cultural authenticity becomes more and more important since it is more and more tenuous. The story or standard of authenticity shifts as ancestors pass, as everyone becomes Americanized and as Italian neighborhoods dwindle and disappear. Often the narrative that the grandparents told, the food and recipes individual or unique to the family or the specific region of Italy, becomes generalized as "authentically" Italian.

It is also important to note that before unification in 1871, the lands of Italy were fragmented villages and city-states (Mangione & Morreale, 1992, p. 31). Political and economic conditions in southern Italy had been horrible for peasants for centuries, and the move toward unification ultimately inspired what would become the mass exodus from the region, many headed for the United States. Italians did not identify themselves as Italians, but rather as former residents of Calabria, Sicily, Naples and/or their surrounding villages or towns. According to many, "Italy" was an invention, and "Italian" was an imposed label. Even while I was growing up in Syracuse, New York in the late twentieth century, saying that you were Italian was not enough, and it was followed up with the question about what *part* of Italy your family came from. Even among the granddaughters of immigrants, these regional distinctions, rooted in stereotypes, provided information about temperament, class and culinary preferences.

The point here is that notions of authenticity are always invented, and this was even the case back in Italy, where family and regional differences, not national identification, dictated the "right" way to do things. Nonetheless, demonstrating that you are a "real" Italian serves a different and weightier purpose for immigrants and their descendants in the US who firmly claim their ethnic identity.

The recognition of other "real" Italians – or, as psychologists would say, group identification – involves the processes of identifying similar others in order to establish where "I" fit and who counts as "us". The meaning and evaluation drawn from group identities result from the ability to make claims about what I "am" as well as what I "am not" and, accordingly, what *we* "are" and what we "are not". Group boundary maintenance and policing results from group identification, especially in relation to what and how much is at stake for its members. Boundaries are fluid, often contradictory and highly contested between and among groups and individuals. Jenkins' (1996) explains, "Identity is about boundary *processes* rather than boundaries. As interactional episodes, those processes are temporary check-points rather than concrete walls" (p. 99).

Claims of authenticity serve as both a policing of boundaries, while also a product of a continuous and shifting process. In her discussion of food at Italian festivals, Magliocco (1993) warns us "not to engage in a fruitless quest for authenticity" because

> the genuineness of a tradition is of no consequence in terms of its importance to the community as a symbol of identity.... Any tradition may be subject to strategic manipulation.... Rather, these elements contain important information about the performer of the item of folklore, its intended audience, and the politic and social relationship between the two. (Magliocco 1993, p. 119)

Method

Participants

Interviewees were 12 "baby boomer" women who self-identified as Italian American living in Syracuse, New York. All are working class and heterosexual. The "baby boom" generally refers to individuals born between 1945 and 1965, making the women in their 1950s and 1960s at the time of interview. Through interview prompts, participants reflected upon recipes and messages received from their parents' generation, and wishes for their descendants. Their life stage enables examination of intergenerational transmission of culture and ethnic identity through food. Focusing on this sample allows for exploration of Italian American femininity in a particular sociohistorical context and geographical location. Women of this era came of age in the 1950s and 1960s, marking a particular moment in American nationalism, while negotiating being the children or grandchildren of Italian immigrants.

All participants were descendants of immigrants, naming one or more grandparents as emigrating from Italy. That is, all interviewees and their parents were born in the US and, for some, several grandparents were born in the US as well, making the women in this study several generations removed from migration. None of the participants could speak Italian, and they were the product of parents and grandparents ensuring the Americanization of their daughters by purposefully not teaching their language to their daughters. In most cases, Italian was reserved for covert communication between grandparents and sometimes parents, but, as children, parents and grandparents excluded the women that we interviewed.

Why Syracuse?

Syracuse is an industrial city in central New York, which derives its name from Siracusa, a city on the eastern coast of Sicily. Italians in Syracuse settled on the North Side of the city, mostly around the turn of the twentieth century, as they did in many other American cities. The significant population of Italian Americans allowed for an ethnographic approach to this study, exploring the ways in which ethnic and immigrant identities are connected to place and space. Syracuse also represents one of the many cities on the east coast of the United States that contains white ethnic enclaves, which is not as omnipresent in other regions such as the west coast of the US. Furthermore, our family has long established roots in Syracuse, and it is where we were born and raised. These connections and relationships to the community facilitated trust in the participants as well as the use of snowball sampling.

The interview

Participants were recruited through family and community relationships, and each woman was asked to bring a recipe to the interview. Participants were informed that they would be asked questions about the recipe of an Italian dish that they brought to the interview, questions about their ideas and feelings about food in general and questions about being Italian American. Interviews averaged about 30 minutes in length, and they were audio recorded and transcribed verbatim.

Other ethnic identity researchers have also employed methodologies that focus participants on an object or cultural product, such as a photograph (Nenga, 2003). For example, Banks (2000) explores how black women's narratives and conversations about hair and hairstyling practices are especially representative of the politics of race, gender, class, identity, self-esteem, social control and power. While Banks might have asked her interviewees questions about these topics directly, discussions about hair seemed to allow women to freely discuss a familiar topic while revealing significant and symbolic meanings. This study is inspired by Banks' work in that the research design was intended to engage Italian American women in discussion about a familiar and tangible topic that is rich with multiple meanings and connections to ethnicity, immigration, gender, generation and class.

Verbatim transcripts were coded for Italian American identity in terms of expressions reflecting *an Italian imaginary*, where participants referenced and imagined and Italian past, indicating partial knowledge or fantasy about what it means to be Italian. Since participants strongly identified as Italian, yet were several generations removed from migration, a constructed notion of Italianness functioned to inform their identities and narratives about food. Similarly, *authenticity* was coded if the participant described indicators of "real" Italians, Italian recipes, food tastes or distinctions between "authentic" Italians and "inauthentic" (or non-Italian) "others". The categories were not mutually exclusive, so a participant might describe both an Italian imaginary and authenticity, and each coded accordingly. Several subcategories emerged as instantiations of these two major themes, which will be described in the results section.

If it's not one thing ... it's your mother

A unique element of this project is that I (Andrea) am working with my mother, Carol, who co-created the interview protocol with me. My mother earned her bachelor's and master's degrees after having her children and building a home with my father, Tony. She is a brilliant and successful reading teacher and taught first grade for many years. I have always thought that she could have been an academic, with her keen ear for language, appetite for knowledge and analytic sensibilities. Carol also still lives in Syracuse, where she, her mother and her grandmother lived all their lives. I left Syracuse in my early twenties for graduate school. Carol's father emigrated from Italy as a young boy, and her family lived among a tight-knit community of Italian-American immigrants and their descendants on "the North side".

Carol was a key recruiter of participants as she is a "baby boomer" herself, and she also conducted a majority of the interviews. It was important that the participants felt they were speaking to a peer who could relate and remember alongside them. While I also have a kind of insider status as an Italian American woman, I also represent the daughter who has left and who became an academic. As this study also centers on research questions about the intergenerational transmission of culture, our relationship mirrors these questions, as mother and daughter contend with our similar and disparate experiences with and relationships to Italian American identity and food.

Because I'm the mother

It seems fitting that the beginning steps of this project on food from an Italian American perspective took place around a kitchen table. At the time, I (Carol) considered the conversation taking place to just be a casual exchange of ideas introduced by my daughter Andrea regarding a topic of interest she wanted to discuss with me and my cousin. It has since been molded and sculpted into a questionnaire and a sharing of recipes encompassing the ideas of what it means to women in upstate New York to be Italian Americans and encompassing how food and traditional recipes play into this identity. My daughter Andrea, whom I consider a genius in her field and generally across the board (she will want to edit this but I will fight her on it), created this project and has given it her expert analysis and interpretation.

My part of this project was to find participants who fit the demographic of living in upstate New York, were women of baby boom age and identified themselves as Italian American. I had easy access to this group because they included many of my family members, longtime friends and significant people in my social circle. I interviewed several participants whom I had never met but who were referred to me by women who had initially been interviewed.

I was hesitant to play such an important role in this project. My daughter, the doctor (PhD), is the expert on qualitative research. I am a retired first grade teacher, completely comfortable conversing with six-year-olds and not at all versed in how to conduct qualitative research. However, with Andrea's encouragement, pep talks and articles mailed to me regarding the process of conducting interviews, I became more confident. I began calling friends and family, requesting their participation. I can honestly say that each time I conducted an interview it always evolved into a truly pleasant experience. With each contact, I began to realize how many emotions, beliefs and revelations are held in the answers to many of the interview questions.

Because the women who were interviewed were my peers, sharing common connections such as approximate age, heritage and background of living our lives in central New York, I believe I understood the sense of ownership they felt about the recipes they shared and the heartfelt memories they recalled during the interviews. During each interview, memories were shared of the people that once gathered around a table but who were no longer alive, had grown to adulthood and moved away or simply did not value the traditional dishes any longer. I understood the sense of longing for what has passed, and perhaps, the women I interviewed responded to our shared experiences. I hope I have contributed to the understanding of the subtle sighs, the quiet seconds when a sense of nostalgia took over, and just how important the memories of family, friends and celebrations shared are held in connection to the traditional recipes shared by each woman interviewed. The food translates into memories.

Results: Constructing Italian American identity

An Italian imaginary

Since most of the participants had never been to Italy themselves and grew up as the grandchildren of Italian immigrants, this category is intended to capture ways in which the women imagine their Italianness. As mentioned previously, although identities are socially constructed or imagined, they are still "real", they should still be taken seriously and they are not experienced as lived realities. Historian Stephanie Coontz devoted her book to the idea of an American romanticization of the "good old days", aptly titled *The Way We Never Were: American Families and the Nostalgia Trap* (1992). The interviews in this chapter are really narratives of Italian *American* identities, a mixture and creation of old-world traditions as interpreted by immigrants nearly a century ago, alongside American ideals, values, ingredients and tastes resulting from a post–World War II culture and economy. Labeling themselves as Italian is situated within an American context, such that even using the language "Italian American" is a bit odd.

"I don't know"

One of the most common ways in which an Italian imaginary was revealed was in response to the question "Is there a more 'authentic' or 'Italian' version of this recipe?" Nine of the twelve interviewees said they did not know. Poignantly, Margaret responded,

> Gee, I don't know. I can't imagine. It tastes just like what I consider authentic tortoni that I've tasted at authentic Italian restaurants.

Not only is she unaware of an alternative, more authentic version of the recipe for tortoni (an ice cream dish), but she cannot *imagine* how she might know. She suggests that it tastes "authentic" to her, so her own palate serves as a standard, alongside "authentic" Italian restaurants. She does not state her reasoning for the authenticity of the restaurants but imagines her own experience to be a measure in and of itself.

Similarly, Linda explains,

> Gee, not that I know of. I've never seen, I watch cooking shows, and I've never seen ... a recipe like this in a cookbook. And I have the original Our Lady of Pompeii cookbook. And I don't think I've seen it in that.

Here, Linda imagines that the standard for authenticity includes cooking shows and a cookbook with recipes from Roman Catholic at the local church. Our Lady of Pompeii was a hub of the Italian American neighborhood while she was growing up, located in what was once the center of the Italian American immigrant community. The contributors of those cookbooks, those who provided the recipes, were either Italian immigrants or their daughters and therefore were seen as "more" Italian, a closer connection to authenticity. Carol, the interviewer, laughs in recognition to hearing the reference.

When discussing her recipe for pasta and beans, Sophia responds,

> I have no idea. I know that my friend Marianne puts crushed tomato in hers.

Here, Sophia is aware that her Italian American friend uses different ingredients in her version of the recipe, but she has "no idea" whose is more "authentic". In each of these instances, the Italian imaginary is rooted in individual respondents' specific family tradition: their family version of the recipe.

"Tradition"

Another common way in which participants expressed an Italian imaginary was by invoking the notion of "tradition". This was most often in response to the questions, "What feelings or emotions are associated with eating this food/dish?" and "What feelings or emotions are associated with making or preparing this food/dish?" Connie replied,

> Oh, it brings me back to my childhood. And I feel, like a connection. Like, okay, Italians were making this for hundreds of years. I do like that. I like that a lot.

Here, she connects the food very specifically to her Italian identity and also to an imaginary Italianness, a "tradition" that the recipe had been made over and over again for "hundreds of years". She also suggests that the *traditional* nature of the recipe brings her an unstated sense of comfort. An especially interesting note is that Connie brought a recipe for "timpano", a layered, Italian version of what might be similar to a shepherd's pie. Connie's version of the recipe included heavy meats and cheeses. While she imagines that the recipe is generations old, it is likely that her ancestors, who emigrated from southern Italy (a region high in poverty), could never have afforded these ingredients in the old country. In fact, many dishes that are known as "typical" Italian food in the United States are quite American inventions. When poor southern Italian immigrants arrived on North America and finally had access to ingredients like meat and cheese, usually reserved for the wealthy landowners back in Italy, they added them to many recipes as a sign of prestige. For example, chicken parmesan, fried and smothered in cheese is part of American, not Italian, cuisine.

When asked about the feelings or emotions associated with the recipe that Paula brought for "cavachione", a fried cookie, she said,

> it would remind me of my childhood. Um, I'd want to eat every last one on the plate. I mean, they're not good for you. I mean, let's face it. They are fried. It's not the healthiest thing for you, but, it's just part of your, what you grew up with. It's part of your tradition."

For Paula, like others, her imagined Italianness is connected to comfort and an immigrant family past. Italian identity is grounded in food, family and an imagined sense of continuity over time and place.

Karen explained the centrality of food in honoring her family's tradition.

> The year my mother passed away, I was determined I was going to do the whole [Christmas] meal she always made.... I made everything and thought I had turned the burner off, and I had turned the burner on high and burned everything I had made.
>
> And I went back to the store and just started all over again. And I can still remember – I just can remember thinking, 'I have to do this. I have to do it for Mom, I have to carry on this tradition.' And my father of course saying, 'Oh, nobody's going to know it's burned. Don't go to all that work again.' And I just was insistent. I had to do it for my Mom.

Karen spoke about her mother's recipe for clams casino, a dish that is commonly served for (southern) Italian American Christmas Eve dinner and that consists entirely of seafood. Earlier in the interview Karen admits that she thinks that there is "probably a more Italian version" of the recipe that would include more breadcrumbs than her mother's version. For Karen, following her mother's recipe is a way of grieving and holding on to a tradition, her family's tradition, which serves as her Italian imaginary.

Family variations in recipes are present in any culture at any time, but for immigrants, the melding of old- and new-world tastes and ingredients, sometimes modified over generations, takes on new meaning in the host country. Moreover, the descendants of immigrants who still strongly identify as Italian, seeing it as a set of regulating ideologies in their own lives, have less and less to mark and signify as part of that tradition. This is especially why food becomes a central hinge of maintaining this identity and one reason why Karen so painfully describes her desperation for preparing her mother's recipe.

Family

As clearly demonstrated by the examples above, another way in which participants expressed an Italian imaginary was by referencing the family. In response to the question, "What feelings or emotions are associated with eating or preparing this food/dish?" 11 out of the 12 women referenced family. Many emotions were named explicitly, mostly joy, happiness and love in remembering the family or sadness and nostalgia in missing large family gatherings and those who have passed. In the example above, Karen's determination about executing the recipe is for her mother, with the sense of duty and obligation of a typical, "traditional" Italian daughter. Two women did not use emotional words at all to respond to the question, but instead they simply discussed the family as group or individual family members. It was almost as if invoking the family stood in to represent emotions, as a way to capture an ideal. For example,

> Carol: What feelings or emotions are associated with eating this food, this dish, for you?
>
> Debbie: It just brings a lot of memories back at the table with mom and dad.
>
> Carol: When you make it, are emotions evoked by your actual preparation of this dish?
>
> Debbie: Just remembering how my mother did this all the time, every week.

Remembering her mother making the food and eating it at the table with her parents serves as a stand-in, a substitution for naming an emotion. The description of the food is her emotional response. Of course, this is not a new idea among Italians. Gambino's (1974) now classic text, aptly entitled *Blood of My Blood,* refers to an Italian phrase often used as a term of familial endearment and solidarity, emphasizing the central role of *la famiglia* in Italian culture. Also in the example mentioned previously, Connie responds to the question about emotion by stating, "And I feel, like a connection." While this is not exactly a "feeling" word, it is implied that this connection is with the tradition and the family, as a foundation of her Italian identity.

Also in answering the question about emotions in relation to her recipe for "Italian meatballs", Donna remembers her father, who took over the cooking responsibilities after her mother died young. She recalls one of her favorite stories about him and his intimate connection to the food and Italian imaginary. Carol and Paula inject this story with great laughter throughout. She describes how her father would make meatballs on Saturday evening in preparation of the Sunday meal. As a teenager, Donna, her siblings and friends would go out on Saturday night and return to Donna's home, with the sauce and meat on the stove. The hungry adolescents would devour the meat, only so her father would awaken on Sunday morning to discover that nearly all the meatballs were gone. She describes the story as one that is told and retold, even at her father's funeral. She says,

> I never knew Italian, but I knew when he was mad 'cause he would start swearing in Italian.... He says, 'everybody ate the meat!' When he passed away, we were in the limo, on the way to the cemetery, and we were talking about that. We were all laughing. And I was telling the guy that drove the limo[, and he] was looking like, 'this guy, this was their father, and they're laughing in the back seat?'

Donna's story reveals several important psychological points. Her fondness is couched in the ways in which her father came to expect that this would happen with his children and their friends. As the father, he could have stopped or forbade it, but the meatball stealing communicates a compliment to his cooking – that the food was so good that they couldn't wait. As a response to the question about emotions related to the recipe, Donna recounts this story that involves her whole family, her father's generosity and her father's performance of outrage, which is also a central feature of Italian American humor and connection. Moreover, the central character in this story is the food, the irresistible meatballs.

Authenticity

This category was coded if the participant described indicators of "real" Italians, Italian recipes, food tastes, distinctions between "authentic" Italians and "inauthentic" (or non-Italian) "others". A term that is important to this discussion and to the Italian American women in this sample is *medigan*. It is Italian American slang for "American". Imagine someone with a thick Italian accent pronouncing the word, and it sounds like "Ah-med-i-gan", which then became truncated to *medigan*. It is never a compliment but always a derogatory term used to mark those who are not Italian. It is used most often as a noun to technically (and disdainfully) describe anyone who is not Italian, but when I think critically about its use, the most usual application is to WASPs (white Anglo-Saxon Protestants) or other "whiter" groups. It is almost never used to describe a person of color or even anyone whose ethnicity is readily identifiable (other than white). This composes a complicated message about race, class, history, ethnicity and culture because it is used to distinguish ourselves from whites as "other" (implying some sort of cultural superiority). It might also be used as patronizing, as in, "the poor *medigans*, they just don't know any better" (implying that "real" Italians do).

The following excerpt is from an interview with Annie for a recipe for stuffed calamari.

Andrea: Do you know if there is a more authentic or more Italian version of the recipe?

Annie: There might be. In Italy, you make things different than we do here. So they might do it without the sauce. I do know that they put the tentacles in. And there are some of us who dont... [points to her husband, who is sitting at the table. He shakes his head, as if to say, "no way!"]

Andrea: so you don't eat the tentacles at all?

Annie: he's half *medigan*, remember? [Laughter].

But I mean, I'm not opposed to eating the tentacles, especially, you know, you can fry them, and whatever. But, no, I don't put them in there, because we have some people in our family... my son-in-law is Slovakian–Czechoslovakian. And the kids, I don't know how the kids would react to that, the grandkids.

Here Annie reveals an Italian imaginary in that she doesn't know if there is a more authentic version of the recipe ("there might be"). She says knows that in Italy they do things differently there, but she doesn't exactly know what that means. She points out that her husband refuses to eat the tentacles, marking him as not a "real" Italian; more specifically (and maybe worse), he is half *medigan*. In this case, his mother was German, which is an especially representative type of non-Italian other, especially since there was also a significant German population in Syracuse. However, Annie *is* a "real" Italian because she would eat the tentacles. Other than her husband, there are other non-Italians that she has to contend with, contributing to her omission of the tentacles in the dish. Her son-in-law is not Italian at all ("Slovakian–Czechoslovakian") and clearly marked as "other". Also, her grandkids will also likely reject such a version of the dish, both because her children have married non-Italians and because the subsequent generations are just more Americanized.

Similarly, Rita admits that when she was married to her husband and was raising children, her cooking "was more *medigan*". When I asked if her husband were Italian, she said,

> He was Irish and Polish.... It's funny because he thought his mom made the best sauce in the world, but it was really like salsa. [Laughs]

> Because they put everything in it. And then he had mine, and he loved it.

She laughs at the ridiculousness of an Italian tomato sauce likened to salsa, and she laughs because her husband didn't know any better but to like it. But once he was exposed to her "authentic" food, he couldn't help but "love it".

"Healthy"

One of the subcategories of this theme of authenticity emerged as participants repeatedly invoked the word "healthy", which they used to reference reasons for ingredients for recipes to be modified, compromising their authenticity or reasons for making and eating the food less frequently than in a previous time. Eight women used this word or made some reference to the idea. In the quote at the beginning of this chapter, Donna laments the possibility of using turkey in a meatball recipe, "Only because of everybody watching the fat content. And trying to eat healthier, you know?" As previously mentioned, Paula also confessed that she would want to devour the fried cookies she discussed: "I mean, let's face it. They are fried. It's not the healthiest thing for you."

Karen minimizes the amount of bacon fat in her mother's recipe for clams casino: "I do cut off as much of the visible fat from the bacon that I can, because we're pretty health conscious, and I typically don't ever eat bacon other than in this recipe." Not only does she cut down on the fat, but she makes a point to explain that she doesn't "ever" eat bacon in anything else, because it is unhealthy. Rita remembers that the general cuisine of her childhood was different from her diet today.

> Everything was fried, and it just wasn't healthy. You know, like they would even use heavy cream in their coffee and things like that.... If I have half-and-half in my coffee, it's fat free.

While she explains the importance of being conscious of fat, heart disease and general health, the authenticity is also reciprocally diminished. Here, women negotiate cooking "Italian" with trying to be health conscious.

Discussion

For the Italian American women in this study, food serves as a symbol of identity, history and belonging. The recipes they shared were from generations gone by. Preparing and eating the food signified the resurrection of their loved ones in some way. As Carol described, "with the food, the essence of them is still there." Recipes also have a voice – of the person who wrote it and who made it, representing a connection to an individual family member. The recipes convey the individual character, personality and intent to capture an experience and tradition through food. The effects of these experiences and of immigration persist after generations of Americanization, highlighting the central role that food plays in identity development and maintenance.

This study has clinical implications for working with children and grandchildren of immigrants from many cultural and ethnic groups. Nearly every culture clings to food as the last vestige or representation of a "homeland", sometimes long forgotten. When treating clients who are immigrants themselves and worried about the diluting of the culture, food will serve to assure them that this is a typical concern and process. Espín (1997) explains how immigrants tend to preserve an idea of the homeland that is fixed in the minds of immigrants in a particular moment in time, with the fantasy that their country is just as they left it. When working with clients who are the descendants of immigrants who are struggling to hold on to a cultural identity in the host country, one might encourage their self-education about their culture of origin, perhaps even reclaiming "traditional" recipes.

Psychologically speaking, food memories are primal, connected to motherhood, very early childhood, feelings of longing, of safety, of care. The recipes that the women spoke of were often intricate and labor intensive, and participants implicitly felt loved by having a mother or grandmother care enough to make this dish. Similarly, the recipes they chose were connected to positive memories – of affection and belonging. We cannot know the recipes that might be lost, forgotten or intentionally discarded if they are associated with negative memories.

Practitioners might also consider implications of this study when treating clients with eating problems (Thompson, 1994). Thompson's (1994) powerful research demonstrated that complicated relationships to food and eating are often connected to systems of power and privilege that extend beyond feminized body image and attractiveness. Alternatively, perhaps practitioners should encourage clients who feel especially isolated to use food in healthy ways to foster a sense of home and community. For example, I (Andrea) never really learned to cook or bake traditional family recipes until I lived away while in graduate school.

Finally, Scott's (2001) notion of the fantasy echo is useful not only in considering social identities in historical contexts but also in understanding our own process of conducting this research, which is still ongoing. The fantasy echo reverberates in reading and rereading the related scholarship and interviews, resonating again in new ways, shifting the lens of our own experiences and reinterpreting both memories and family stories. Similarly, the therapeutic process often takes on its own fantasy echo, as we visit and revisit familiar themes in our life histories, some of which belong to us and some of which are the legacies of our families, sometimes connected to migration in ways we never know. As the women in our study continue to navigate the "eating of Italy", their daughters and granddaughters will inherit these legacies in the making of America.

References

Banks, I. (2000). *Hair matters: Beauty, power, and black women's consciousness.* New York: New York University Press.

Bona, M. (1999). *Claiming a tradition: Italian American women writers.* Carbondale, IL: Southern Illinois University Press.

Combahee River Collective (1977/2005). A Black feminist statement. In W.K. Kolmar and F. Bartkowski (Eds.), *Feminist theory: A reader* (2nd ed.) (pp. 311–316). Boston: McGraw Hill.

Coontz, S. (1992). *The way we never were: American families and the nostalgia trap.* New York: Basic Books.

Crenshaw, K.W. (1995). Mapping the margins: Intersectionality, identity politics, and violence against women of color. In K.W. Crenshaw, N. Gotanda, G. Peller & K. Thomas (Eds.), *Critical race theory: The key writings that formed the movement* (pp. 359–383). New York: The New Press.

DeSalvo, L. (2004). *Crazy in the kitchen: Food, feuds, and forgiveness in an Italian American family.* New York: Bloomsbury.

DeSalvo L. & Giunta, E. (Eds.) (2002). *The milk of almonds: Italian American women writers on food an culture.* New York: The Feminist Press.

Dottolo, A.L. & Stewart, A.J. (2008). "Don't ever forget now, you're a Black man in America": Intersections of race, class and gender in encounters with the police. *Sex Roles, 59*, pp. 350–364.

Espín, O.M. (1997). *Latina realities: Essays on healing, migration, and sexuality.* Boulder, CO: Westview Press.

Gabaccia, D. & Iacovetta, F. (2002). *Women, gender, and transnational lives: Italian workers of the world.* Toronto: University of Toronto Press.

Gambino, R. (1974). *Blood of my blood: The dilemma of the Italian-Americans.* New York: Anchor Books.

Giunta, E. (2002). *Writing with an accent: Contemporary Italian American women authors.* New York: Palgrave.

Guglielmo, T.A. (2003). *White on arrival: Italians, race, color and power in Chicago, 1890–1945.* New York: Oxford University Press.

Guglielmo, J. (2010). *Living the revolution: Italian women's resistance and radicalism in New York City, 1880–1945.* Columbia, SC: The University of North Carolina Press.

Jenkins, R. (1996). *Social identity*. New York: Routledge.

Magliocco, Sabina. (1993). Playing with food: The negotiation of identity in the ethnic display event by Italian Americans in Clinton, Indiana. In L. delGiudice (Ed.), *Studies in Italian American Folklore* (pp. 107–126). Logan, UT: Utah State University Press.

Mangione, J. & Morreale, B. (1992). *La storia: Five centuries of the Italian American experience*. New York: Harper Perennial.

Marecek, J. (1995). Gender, politics, and psychology's ways of knowing. *American Psychologist, 50*(3), pp. 162–163.

Nenga, S.K. (2003). Social class and structures of feeling in women's childhood memories of clothing, food, and leisure. *Journal of Contemporary Ethnography, 32*(2), pp. 167–199.

Scott, J.W. (2001). Fantasy echo: History and the construction of identity. *Critical Inquiry, 27*, pp. 284–304.

Tajfel, H. (1981). *Human groups and social categories: Studies in social psychology*. Cambridge: Cambridge University Press.

Thompson, B. (1994/2001). "A Way outa no way": Eating problems among African-American, Latina, and white women. In L. Richardson, V. Taylor & N. Whittier (Eds.), *Feminist frontiers, 5* (pp. 413–421). New York: McGraw Hill.

Conclusion

Oliva M. Espín and Andrea L. Dottolo

This book is the product of years of collaboration and friendship between the two of us. It is the culmination of many hours of sharing mutual concerns that go beyond the creation of this collection. Editing this volume has given both of us the opportunity to discuss in depth many issues that are dear to our hearts. Our professional concerns and expertise are reflected in our attempt to respond to many questions in the psychology of immigrant women.

In this volume, we aimed to present the multifarious problems and possibilities of the experience of migration for women and girls. Despite the unavoidable limitations created by the material we received in response to our request for contributions, the book covers an expanse of topics. The authors have contributed a variety of perspectives and approaches confronted by immigrant women. In addition to the importance of this scholarly enterprise, these chapters reflect the lived experiences of immigrant women. Many of the authors present their professional expertise as well as the story of their own lives. This is not surprising, since some of the authors are immigrants themselves or have been deeply affected by the experiences of their immigrant ancestors.

What it means to be *from* a place and to go *to* a place has fluctuating implications for citizenship, identity and both physical and psychological survival.

We know that immigrants are "betwixt and between" in many areas of their lives, which is especially true for immigrant women and girls. However, this is not necessarily always detrimental, because in their sharing more than one cultural heritage they participate in many enriching experiences. Moreover, this "between-ness" transforms immigrants into cultural bridges, enriching not only their own lives but also the society in which they now live (Phan, 2003). Immigrant women "live and move and have their being in the interstice between the [main stream] American culture and their own" (Phan, 2003, p. 237). While it is true that in modern society many individuals find themselves in situations of multiple belongings, for immigrants this is an existential condition that demands constant negotiation of identities. "Belonging to several worlds and cultures, immigrants have the opportunity to fuse them together and ... fashion a new and different world. ... Thus, being betwixt and between can bring about personal and societal transformation and enrichment" (Phan, 2003, p. 235). We hope this book shows that, despite the unavoidable struggles, immigrant women can, need to and do find ways of integrating their worlds together in order to live healthy lives.

As the chapters in this book reveal, the reality of being "betwixt and between" in many daily circumstances demands psychological resources. For women and girls, gender role expectations in their home cultures that may contradict the expectations of the host society pose an additional task. As we have seen in these chapters, immigrant women's gender expressions and identities as well as sexuality and behaviors are not static. They are involved in forms of cultural change in their everyday choices. "In cultural contexts in which sexist beliefs are considered traditional, shifts toward gender equality represent an example of cultural change" (Rosenthal, Levy & Militano, 2014, p. 519). As the authors of the chapters in this book show us, even slight shifts in belief and behavior contribute to the transformation of the home culture and, in more or less subtle ways, also impact the host culture. For immigrant women and girls, US culture and immigrant cultures are brought daily into "reciprocal engagement in such a way that both of them are transformed from within. ... The expressions of both cultures are transformed as the result of this process" (Phan, 2003, p. 238).

Despite the richness and variety of the material included, we are aware that there are some topics that are absent from this volume. For example, the focus of the chapters, with one exception, refers to immigration to the United States. We wish we would have had more in-depth discussions and a more complicated class analysis that would have included issues of class in the context of migration for both home and host societies. Social class is a powerful factor in decisions about migration and the capacity to do it, opportunities afforded to immigrants in the home and host countries, everyday life attitudes and a myriad of details that influence the lives of immigrants and their descendants. A class analysis (versus a mere mentioning or recognition of social class) not only includes an examination of financial status (hence, we avoid the term "socioeconomic status") but also includes an exploration of cultural capital and relationships to political and institutional structures. Attention to social class as an explicit part of intersectionality as both a methodological and theoretical framework can only strengthen and deepen our understanding (Dottolo & Stewart, 2008).

We also would have wanted to discuss issues of sexuality more in depth because we believe that these are extremely significant for the process of acculturation in immigrant women (e.g. Espín, 1997, 1999). Particularly absent from the book are questions raised by the experiences of lesbian immigrants. Lesbians, gay men, transgender individuals and others who identify as queer obviously exist among immigrants, although many immigrant communities prefer to believe otherwise. Some posit that homosexuality is just an "evil" of modern societies, acquired by "contagion" or through the bad influences of the new society (e.g. Espín, 1987; 1996; 2012; Tremble, Schneider & Appathurai, 1989). Gender identity issues and homosexuality are considered to be one and the same manifestation of those evils of modern American society that overcome children in immigrant families.

Difficulties prevail in obtaining adequate statistics on sexually nonconforming populations among immigrants and refugees. Most likely, several million lesbians and transgender women are among the immigrant population in the United States. The numbers are even higher if we factor immigrants and refugees worldwide.

The emotional and practical difficulties of migration are compounded for them by the difficulties created by prevalent conceptions of sexuality and gender in both the home and the host cultures, familial and societal rejection of their sexual and gender identity and legal restrictions on their immigration. Still, the new environment may open up possibilities hitherto unavailable in the country of birth. Paramount among these is the possibility of living a more authentic life, which is not to say that the United States provides a warm welcome for lesbians, gay men or transgender individuals.

Legislation concerning the immigration of sexually nonconforming individuals into the United States has been mostly restrictive. The Immigration Act of 1917 excluded "persons of constitutional psychopathic inferiority". In 1952 the Immigration and Nationality Act (INA) reworded its exclusions: "aliens afflicted with psychopathic personality, epilepsy or a mental defect" were not allowed to enter the United States. The legislative history of these acts shows that this language was broad enough to exclude "homosexuals and sex perverts". In 1952 the INA was adopted in a climate of national paranoia. At that time, homosexuality was associated with communism, not only in the popular mind but by the government as well. Ironically, the Communist Party, like mainstream culture, rejected homosexuals as well. Then, as more recently with HIV/AIDS, migrants continued to be considered culprits in health epidemics (Manalansan, 2006, p. 228).

These laws started changing in 1979. As a consequence of the American Psychiatric Association's 1973 declaration that homosexuality per se does not constitute mental illness, the Surgeon General announced that the Public Health Service, working then under the Immigration and Naturalization Service (INS), would no longer exclude aliens suspected of homosexuality. In 1980, the same federal agency adopted new procedures devoid of questions about sexual orientation. Finally, the Immigration Act of 1990 abandoned these provisions. However, "the enormous gap between removing explicit discrimination from the law and ensuring that equal access can be realized in practice" (Luibhéid, 2005, p. xiii) continues to be present.

Yet, despite the historically negative reception or outright rejection manifest through US immigration laws, in many cases immigration to the US is still considered a positive alternative to the harsher persecution some may confront in their homeland.

One area that remains problematic is the issue of same-sex marriage. While immigration laws have traditionally permitted the immigration of a foreign spouse, lesbians and gay men have been prevented from bringing foreign partners to the US because same–sex marriages have not been considered legal. This is an active area of transformation at this point in history. The recent Supreme Court decision about the Defense of Marriage Act (DOMA) has radically changed this situation. Its consequences for the family life of non-heterosexual immigrants are beginning to be felt.

Sexuality, although a universal component of human experience, is embodied and expressed in various ways. Even the definition of what is "sexual" changes among cultures. These cultural constructs are intertwined with lesbians' expressions of their sexuality. As lesbian communities also vary across countries and even regionally within the same country, many women who identified as lesbian or transgender before the migration have to "learn" to be women and lesbians in their new cultural context. If they come from a non–European background, they also face acculturation as persons of color. Many of these women have experienced discrimination because of their sexuality before the migration. Some experience discrimination based on sexuality combined with ethnic discrimination after migration. As I found when researching the identity development of immigrant Latina lesbians in the 1980s,

> Coming out to self and others in the context of a sexist and heterosexist American society is compounded by coming out in the context of what is usually a sexist and heterosexist [culture of origin] immersed in racist society. (Espín, 1987, p. 35)

Every person who doesn't sexually conform, regardless of ethnicity, must resolve a number of issues, such as deciding whether or not to disclose to the family and/or to strangers, finding a niche among peers and reconciling sexual and gender identities with other aspects of identity. For the child of immigrant parents, the process of identity development takes place against the backdrop of ethnic traditions, values and social networks. The rift that occurs between parent and child over sexuality and identity is set in the context of an existing conflict as the child pulls away from the home culture to espouse the host culture (Tremble, Schneider & Appathurai, 1989, p. 255). In other words, young lesbians encounter additional struggles while developing self-understanding and identity in the context of a hostile world (Pendragon & Espín, 2010). We wish these could have been better represented in this book.

By the same token, issues of disability or of age and ageism are absent from these chapters, even though some discuss the different experiences encountered by immigrant women of different ages.

But we are convinced that this is only the beginning of a conversation that needs to be extended in psychology and gender studies. And since gender is a lens for the analysis of power differentials, we also need a more pointed interrogation of the construction of masculinities in the context of migration. Along those same lines, we urge scholars to consider a more sophisticated unpacking of privilege in the context of migration, not only of masculinities, but of whiteness, heterosexualities, able-bodiedness and upper-middle-class backgrounds, to name a few. When we imagine that certain identities or social categories are relevant to only those who are marginalized or oppressed, we maintain our attention on those who are disadvantaged by social systems instead of also interrogating those who benefit and are in positions of power to make change. We agree with Brekhus (1998) that a "reverse marking" of privileged identities is necessary in order to make the subject of scrutiny those categories that are hidden, invisible and considered "normal" (Dottolo, 2014; Dottolo & Stewart, 2013). Being born in the US versus being "foreign born" is one such category of privilege affecting the life situation of millions of people in this country.

In a contemporary and future global context, geographic and political landscapes are changing faster than they ever have before. At the time of this writing, national and regional borders are shifting in places such as the Middle East, North Africa and Eastern Europe. Climate change and global warming are reconstructing the landscape, eroding and potentially erasing regions and entire nations. The pace and context of this change presents a unique and significant opportunity for scholars and practitioners of migration.

We hope the chapters from this volume inspire future research and policy questions, continuously demonstrating the imperative of understanding gender and integrating intersectionality as part of the methodologies of acquiring answers.

The contributions of this volume demonstrate the urgency for a feminist psychology of migration, one that explores and provides a feminist understanding of how migration shapes how not only immigrants but also their communities, home countries and host cultures can think, feel and know. We hope that this collection contributes to present and future discourses on the gendered psychology of migration, and we are excited to witness transformed and transforming gendered journeys yet to come.

References

Brekhus, W. (1998). A sociology of the unmarked: Redirecting our focus. *Sociological Theory, 16*(1), pp. 34–51.

Dottolo, A.L. (Guest editor) (2014). Special issue: Lesbians and white privilege. *Journal of Lesbian Studies, 18*(2). Retrieved from http://www.tandfonline.com/eprint/JG3Ib5PmDx3TAnDvpaBy/full.

Dottolo, A.L. & Stewart, A.J. (2013). "I never think about my race": Psychological features of white racial identities. *Qualitative Research in Psychology, 10*, pp. 102–117. doi:10.1080/14780887.2011.586449.

Dottolo, A.L. & Stewart, A.J. (2008). "Don't ever forget now, you're a Black man in America": Intersections of race, class and gender in encounters with the police. *Sex Roles, 59*, pp. 350–364.

Espín, O.M. (1987). Issues of identity in the psychology of Latina lesbian women. In *Lesbian Psychologies Collective* (Eds.) (pp. 35–55) Urbana: University of Illinois Press.

Espín, O.M. (1996). Immigrant and refugee lesbians. In E.D. Rothblum & L.A. Bond (Eds.), *Preventing heterosexism and homophobia* (pp.174–183). Thousand Oaks, CA: Sage.

Espín, O.M. (1997). *Latina realities: Essays on healing, migration and sexuality.* Boulder, CO: Westview Press.

Espín, O.M. (1999). *Women crossing boundaries: A psychology of immigration and transformation of sexuality*. New York: Routledge.
Espín, O.M. (2012). "An illness we catch from american women"? The multiple identities of Latina Lesbians. *Women & Therapy, 35*(1/2), pp. 45–56.
di Leonardo, M. (1987). The female world of cards and holidays: Women, families, and the work of kinship. Signs: *Journal of Women in Culture and Society, 12*(3), pp. 440–453.
Luibhéid, E. (2005). Introduction: Queering migration and citizenship. In L. Cantú & E. Luibhéid (Eds.), *Queer migrations: Sexuality, U.S. citizenship, and border crossings* (pp. ix–xlvi). Minneapolis, MN: University of Minnesota Press.
Manalansan, M.F. (2006). Queer intersections: Sexuality and gender in migration studies. *International Migration Review, 40*(1), pp. 224–249.
Phan, P.C. (2003). *Christianity with an Asian face: Asian American theology in the making*. New York: Orbis Books.
Pendragon, D.K. & Espín, O.M. (Eds.) (2010). *Out There! Exploring the Varied Experiences of Lesbian Youth* (Special Issue of *Journal of Lesbian Studies*). New York: Taylor and Francis.
Rosenthal, L., Levy, S.R. & Militano, M. (2014). Polyculturalism and sexist attitudes: Believing cultures are dynamic relates to lower sexism. *Psychology of Women Quarterly, 38*(4), pp. 519–534.
Tremble, B., Schneider, M. & Appathurai, C. (1989). Growing up gay or lesbian in multicultural context. In G. Herdt (Ed.), *Gay and lesbian youth* (pp. 253–267). New York: Harrington Park Press.

Index

CPSIA information can be obtained
at www.ICGtesting.com
Printed in the USA
LVOW01*1013131216
517056LV00009B/63/P

9 781137 521460